Multivariable Calculus and *Mathematica*®

With Applications to Geometry and Physics

Multivariable Calculus and *Mathematica*®

With Applications to Geometry and Physics

Kevin R. Coombes
Ronald L. Lipsman
Jonathan M. Rosenberg

Kevin R. Coombes
Ronald L. Lipsman
Jonathan M. Rosenberg
Department of Mathematics
University of Maryland
College Park, MD 20472-4015
USA

Library of Congress Cataloging-in-Publication Data
Coombes, Kevin Robert, 1955–
    Multivariable calculus and Mathematica : with applications to
geometry and physics / by Kevin R. Coombes, Ronald L. Lipsman, Jonathan
M. Rosenberg
        p.   cm.
    Includes bibliographical references and index.
    ISBN 0-387-98360-0 (pbk. : alk. paper)
    1. Calculus—Computer-assisted instruction.   2. Mathematica
(Computer file).   I. Lipsman, Ronald L.   II. Rosenberg, J.
(Jonathan), 1951–   .   III. Title.
QA303.5.C65C66   1998
515'.84'078553042—dc21   97-44764

Printed on acid-free paper.

Mathematica is a registered trademark of Wolfram Research, Inc.

Production managed by Anthony K. Guardiola; manufacturing supervised by Jeffrey Taub.
Photocomposed copy prepared using the authors' TeX files.
Printed and bound by Hamilton Printing Co., Rensselaer, NY.
Printed in the United States of America.

9 8 7 6 5 4 3 2 1

ISBN 0-387-98360-0 Springer-Verlag New York Berlin Heidelberg   SPIN 10647692

TELOS, The Electronic Library of Science, is an imprint of Springer-Verlag New York. Its publishing program encompasses the natural and physical sciences, computer science, economics, mathematics, and engineering. All TELOS publications have a computational orientation to them, as TELOS' primary publishing strategy is to wed the traditional print medium with the emerging new electronic media in order to provide the reader with a truly interactive multimedia information environment. To achieve this, every TELOS publication delivered on paper has an associated electronic component. This can take the form of book/diskette combinations, book/CD-ROM packages, books delivered via networks, electronic journals, newsletters, plus a multitude of other exciting possibilities. Since TELOS is not committed to any one technology, any delivery medium can be considered.

The range of TELOS publications extends from research level reference works through textbook materials for the higher education audience, practical handbooks for working professionals, as well as more broadly accessible science, computer science, and high technology trade publications. Many TELOS publications are interdisciplinary in nature, and most are targeted for the individual buyer, which dictates that TELOS publications be priced accordingly.

Of the numerous definitions of the Greek word "telos," the one most representative of our publishing philosophy is "to turn," or "turning point." We perceive the establishment of the TELOS publishing program to be a significant step towards attaining a new plateau of high quality information packaging and dissemination in the interactive learning environment of the future. TELOS welcomes you to join us in the exploration and development of this frontier as a reader and user, an author, editor, consultant, strategic partner, or in whatever other capacity might be appropriate.

TELOS, The Electronic Library of Science
Springer-Verlag New York

TELOS Diskettes

Unless otherwise designated, computer diskettes packaged with TELOS publications are 3.5" high-density DOS-formatted diskettes. They may be read by any IBM-compatible computer running DOS or Windows. They may also be read by computers running NEXTSTEP, by most UNIX machines, and by Macintosh computers using a file exchange utility.

In those cases where the diskettes require the availability of specific software programs in order to run them, or to take full advantage of their capabilities, then the specific requirements regarding these software packages will be indicated.

TELOS CD-ROM Discs

For buyers of TELOS publications containing CD-ROM discs, or in those cases where the product is a stand-alone CD-ROM, it is always indicated on which specific platform, or platforms, the disc is designed to run. For example, Macintosh only; Windows only; cross-platform, and so forth.

TELOSpub.com (Online)

Interact with TELOS online via the Internet by setting your World-Wide-Web browser to the URL: *http://www.telospub.com*.

The TELOS Web site features new product information and updates, an online catalog and ordering, samples from our publications, information about TELOS, data-files related to and enhancements of our products, and a broad selection of other unique features. Presented in hypertext format with rich graphics, it's your best way to discover what's new at TELOS.

TELOS also maintains these additional Internet resources:

gopher://gopher.telospub.com
ftp://ftp.telospub.com

For up-to-date information regarding TELOS online services, send the one-line e-mail message:

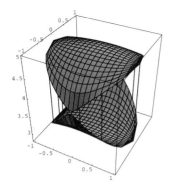

Preface

The preface of a book gives the authors their best chance to answer an extremely important question: What makes this book special?

This book is our attempt to enrich and enliven the teaching of multivariable calculus and mathematical methods courses for scientists and engineers. Most books in these subjects are not substantially different from those of fifty years ago. (Well, they may include fancier graphics and omit several topics, but those are minor changes.) This book is different. We do touch on most of the classical topics; however, we have made a particular effort to illustrate each point with a significant example. More importantly, we have tried to bring fundamental physical applications—Kepler's laws, electromagnetism, fluid flow, energy estimation— back to a prominent position in the subject. From one perspective, the subject of multivariable calculus only exists because it can be applied to important problems in science.

In addition, we have included a discussion of the geometric invariants of curves and surfaces, providing, in effect, a brief introduction to differential geometry. This material provides a natural extension to the traditional syllabus.

We believe that we have succeeded (in resurrecting material that used to be in the course while introducing new material) for one simple reason: we use the computational power of the mathematical software system *Mathematica* to carry a large share of the load. *Mathematica* is tightly integrated into every portion of this book. We use its graphical capabilities to draw pictures of curves and surfaces; we use its symbolical capabilities to compute curvature and torsion; we use its numerical capabilities to tackle problems that are well beyond the typical mundane examples of textbooks that treat the subject without using a computer. As an added convenience for the user, we have included with this book a computer disk, where the *Mathematica* programs used in the text are reproduced.

As an additional benefit from introducing *Mathematica*, we are able to improve students' understanding of important elements of the traditional syllabus. Our students are better able to visualize regions in the plane and in space. They develop a better feel for the geometric meaning of the gradient; for the method of steepest descent; for the orthogonality of level curves and gradient flows. Because they have tools for visualizing cross sections of solids, they are better able to find the limits of integration in multiple integrals.

To summarize, we think this book is special because, by using it:

- students obtain a better understanding of the traditional material;
- students see the deep connections between mathematics and science;
- students learn more about the intrinsic geometry of curves and surfaces;
- students acquire skill using *Mathematica*, a powerful piece of modern software; and
- instructors can choose from a more exciting variety of problems than in standard textbooks.

Conventions

Throughout the book, *Mathematica* commands, such as `Solve`, are printed in typewriter boldface. Menu options, such as `Cell`, are printed in a monospaced typewriter font. Theorems and general principles, such as: *derivatives measure change*, are printed in a slanted font. When new terms, such as *torsion*, are introduced, they are printed in an italic font. Everything else is printed in a standard font.

Each chapter and its accompanying Problem Set have a small illustration in the upper left-hand corner of the first page. Each figure is taken from our *Mathematica* solution to one of the problems in the Problem Set. We leave it to the industrious reader to determine which problems give rise to the figures.

Acknowledgments

The authors are pleased to acknowledge that some of the ideas for problems in Problem Sets C and F were obtained from *Modern Differential Geometry of Curves and Surfaces* by A. Gray (CRC Press, 1993), and from *Elementary Differential Geometry* by B. O'Neill (Academic Press, 1966).

We thank several of our colleagues who helped test preliminary versions of this book in their classrooms: Der-Chen Chang, Paul Green, Oscar Gonzalez, Alessandra Iozzi, Jonathan Poritz, and Garrett Stuck. We also thank Alvaro Alvarez-Parrilla and Nic Ormes, who served capably as teaching assistants. Finally, we thank four TELOS reviewers for their very helpful comments on the manuscript: William Barker of Bowdoin College, Al Hibbard of Center College, Iowa, Jamie Radcliffe of the University of Nebraska–Lincoln, and Todd Young, formerly at Northwestern University and now at Ohio University.

Kevin R. Coombes
Ronald L. Lipsman
Jonathan M. Rosenberg
College Park, Maryland

Contents

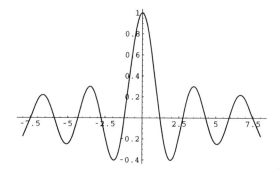

Chapter 1

INTRODUCTION

$\mathbf{W}$e wrote this book as a supplement for the third semester of a physical science or engineering calculus sequence. It can equally well be used in a postcalculus course or problem seminar on mathematical methods for scientists and engineers. The subject is traditionally called *Calculus of Several Variables*, *Vector Calculus*, or *Multivariable Calculus*. The usual content is:

- *Preliminary Theory of Vectors*: Dot and Cross Products; Vectors, Lines, and Planes in $\mathbf{R}^3$.
- *Vector-Valued Functions*: Derivatives and Integrals of Vector-Valued Functions of One Variable; Space Curves; Tangents and Normals; Arclength and Curvature.
- *Partial Derivatives*: Directional Derivatives; Gradients; Surfaces; Tangent Planes; Multivariable Max/Min Problems; Lagrange Multipliers.
- *Multiple Integrals*: Double and Triple Integrals; Cylindrical and Spherical Coordinates; Change of Variables.
- *Calculus of Vector Fields*: Line and Surface Integrals; Fundamental Theorem of Line Integrals; Green's, Stokes', and Divergence Theorems.

We have tried to modernize the course, in part by introducing the mathematical software system *Mathematica* as a powerful tool. Here are our reasons:

- to remove the drudgery from tedious hand calculations that can now be done easily by computer;

- to improve students' understanding of fundamental concepts in the traditional syllabus; and
- to introduce new geometrical and physical topics.

Benefits of Mathematical Software

To elaborate, we describe some benefits that follow from introducing *Mathematica*. First, the traditional multivariable calculus course has a tremendous geometric component. Students struggle to handle it. Unless they are endowed with artistic gifts or uncanny geometric insight, they may fail to depict and understand the geometric constructs. Often, they rely on illustrations in their text or prepared by their instructor. While the quality of those illustrations may be superior to what they can generate themselves, spoon-fed instruction does not lead to the same depth of understanding as self-discovery. Providing a software system like *Mathematica* enables all students to draw, manipulate, and analyze the geometric shapes of multivariable calculus.

Second, most of the numbers, formulas, and equations found in standard problems are highly contrived to make the computations tractable. This places an enormous limitation on the faculty member trying to present meaningful applications, and lends an air of untruthfulness to the course. (Think about the limited number of examples for an arclength integral that can be integrated easily in closed form.) With the introduction of *Mathematica*, this drawback is ameliorated. The numerical and symbolic power of *Mathematica* greatly expands our ability to present realistic examples and applications.

Third, the instructor can concentrate on nonrote aspects of the course. The student can rely on *Mathematica* to carry out the mundane algebra and calculus that often absorbed all of the student's attention previously. The instructor can focus on theory and problems that emphasize analysis, interpretation, and creative skills. Students can do more than crank out numbers and pictures; they can learn to explain coherently what they mean. This capability is enhanced by the *Mathematica* Notebook interface, which allows the student to integrate *Mathematica* commands with output, graphics, and textual commentary.

Fourth, the instructor has time to introduce modern, meaningful subject matter into the course. Because we can rely on *Mathematica* to carry out the computations, we are free to emphasize the ideas. In this book, we concentrate on aspects of geometry and physics that are truly germane to the study of multivariable calculus. With the introduction of *Mathematica*, this material can, for the first time, be presented effectively at the sophomore level.

What's in This Book

The bulk of the book consists of eight chapters (numbered 2–9) on multivariable calculus and its applications. Some chapters cover standard material from a non-standard point of view; others discuss topics that are hard to address without using a computer.

Each chapter is accompanied by a problem set. The problem sets constitute an integral part of the book. Solving the problems will expose you to the geometric, symbolic, and numerical features of multivariable calculus. Many of the problems (especially in Problem Sets C–I) are not routine.

Each problem set concludes with a Glossary of *Mathematica* commands, accompanied by a brief description, which are likely to be useful in solving the problems in that set. A more complete Glossary, with examples of how to use the commands, is included at the back of the book. In addition, the book contains *Mathematica* Tips, Sample Notebook Solutions, and an Index. Finally, the accompanying disk contains:

- electronic versions of the Sample Notebook Solutions;
- *Mathematica* Notebooks for each chapter, containing the *Mathematica* input lines that recreate all of the output and figures from that chapter; and
- a *Mathematica* Notebook containing all the sample input lines from the Glossary.

Chapter Descriptions

Chapter 1, *Introduction*, and Problem Set A, *Review of One-Variable Calculus*, describe the purpose of the book and its prerequisites. The Problem Set reviews both the elementary *Mathematica* commands and the fundamental concepts of one-variable calculus needed to use *Mathematica* to study multivariable calculus.

Chapter 2, *Vectors and Graphics*, and Problem Set B, *Vectors and Graphics*, introduce the mathematical idea of vectors in the plane and in space. We explain how to work with vectors in *Mathematica* and how to graph curves and surfaces in space.

Chapter 3, *Geometry of Curves*, and Problem Set C, *Curves*, examine parametric curves, with an emphasis on geometric invariants like speed, curvature, and torsion, which can be used to study and characterize the nature of different curves.

Chapter 4, *Kinematics*, and Problem Set D, *Kinematics*, apply the theory of curves to the physical problems of moving particles and planets.

Chapter 5, *Directional Derivatives*, and Problem Set E, *Directional Derivatives and the Gradient*, introduce the differential calculus of functions of several variables, including partial derivatives, directional derivatives, and gradients. We also explain how to graph functions and their level curves or surfaces with *Mathematica*.

Chapter 6, *Geometry of Surfaces*, and Problem Set F, *Surfaces*, study parametric surfaces, with an emphasis on geometric invariants, including several forms of curvature, which can be used to characterize the nature of different surfaces.

Chapter 7, *Optimization in Several Variables*, and Problem Set G, *Optimization*, discuss how calculus can be used to develop numerical algorithms. We also explain how to use *Mathematica* to test and apply these algorithms in concrete problems.

Chapter 8, *Multiple Integrals*, and Problem Set H, *Multiple Integrals*, develop the integral calculus of functions of several variables. We show how to use *Mathematica* to set up multiple integrals, as well as how to evaluate them.

Chapter 9, *Physical Applications of Vector Calculus*, and Problem Set I, *Physical Applications*, develop the theories of gravitation, electromagnetism, and fluid flow, and then use them with *Mathematica* to solve concrete problems of practical interest.

Chapter 10, *Mathematica Tips*, gathers together the answers to many *Mathematica* questions that have puzzled our students. Read through this chapter at various times as you work through the rest of the book. If necessary, refer back to it when some aspect of *Mathematica* has you stumped.

The *Glossary* includes all the commands from the Problem Set glossaries—together with illustrative examples—plus some additional entries.

The *Sample Notebook Solutions* contain sample solutions to one or more problems from each Problem Set. These samples can serve as models when you are working out your own solutions to other problems.

Finally, we have a comprehensive *Index* of *Mathematica* commands and mathematical concepts that are found in this book.

What's Not in This Book

This book is not a self-contained introduction to multivariable calculus; it was written to supplement a standard (or "reformed") textbook. We have included the portions of mathematics and physics that we find the most interesting, and have freely omitted routine matters that are easily obtained elsewhere. Although it is theoretically possible for a dedicated instructor to teach a course in multivariable calculus using this book as the sole text, none of the authors has (yet) done so.

Required *Mathematica* Background

Neither is this book a self-contained introduction to *Mathematica*. We assume that you can learn the basics of *Mathematica* elsewhere. Since we cannot refer to a "standard text" for this purpose, here is a detailed description of prerequisites.

We assume that you know how to start *Mathematica* on your computer. We also assume that you are familiar with elementary *Mathematica* commands to do arithmetic, algebra, and one-variable calculus. More precisely, we assume that you can `Factor` and `Simplify` algebraic expressions, `Solve` equations, and differentiate, integrate, and compute limits (with the *Mathematica* commands `D`, `Integrate`, and `Limit`). Finally, we assume that you can `Plot` functions of one variable. Problem Set A tests exactly this body of knowledge in *Mathematica*.

We also assume that you can use *Mathematica* Notebooks—in particular, that you can combine input, output, graphics, and text to produce a coherent and attractive document. You can learn how to use *Mathematica* Notebooks from any of the primers cited below. You may also consult *The Mathematica Book*, Third Edition, by Stephen Wolfram (Wolfram Media, Inc. and Cambridge University Press, 1996), but that is a reference book of far greater scope than is required here.

If necessary, you can quickly gain expertise in *Mathematica* from the online Help Browser or from one of the following primers:

- *Mathematica by Example*, Second Edition, by M. L. Abell and J. P. Braselton (AP Professional, 1997);
- *Mathematica: A Practical Approach*, Second Edition, by N. Blachman and C. P. Williams (Prentice Hall, 1997);
- *First Steps in Mathematica*, by W. Burkhardt (Springer-Verlag, 1994);
- *The Mathematica Primer*, by K. R. Coombes, B. R. Hunt, R. L. Lipsman, J. E. Osborn, and G. J. Stuck (Cambridge University Press, 1997);
- *A Tutorial Introduction to Mathematica*, by W. Ellis, Jr. and E. Lodi (Brooks/Cole, 1991); and
- *Beginner's Guide to Mathematica Version 3*, by T. Gray and J. Glynn (Cambridge University Press, 1997).

More *Mathematica* books are being published all the time. You can find an up-to-date list in the bookstore section of the Wolfram Research web site at

$$\texttt{http://www.wolfram.com}$$

How to Use This Book

If, as in our courses, this book supplements a traditional text, then it contains more material than can be covered in a single semester. To aid in selecting a coherent subset of the material, here is a diagram showing the dependence among the chapters.

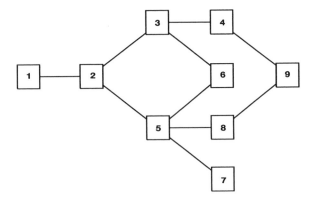

We suggest that you work all the problems in Problem Set A, read Chapter 2, and then work at least a quarter to a half of Problem Set B. After that, various combinations of chapters are possible. Here are a few selections that we have found suitable for a one-semester multivariable calculus course:

- *Geometry Emphasis:* Chapters 3, 5, and 6 with Problem Sets C, E, and F. If time permits, you could include portions of Chapter 4 and Problem Set D, or Chapter 8 and Problem Set H.
- *Physical Applications Emphasis:* Chapters 3, 4, and 9 with Problem Sets C, D, and I. If time permits, it is desirable to add parts of Chapters 5 and 8 with Problem Sets E and H.
- *Calculus Emphasis:* Chapters 5, 7, and 8 with Problem Sets E, G, and H. If time permits, it is desirable to include portions of Chapter 3 and Problem Set C.

In a problem seminar or mathematical methods course, more flexibility is possible, and we could choose a greater variety of problems from various chapters.

Beginning with Problem Set C, the exercises in the Problem Sets become fairly substantial; it is easy to spend an hour on each problem. To ease the burden, we often allow students to collaborate on the problems in groups of two or three. We ask each collaborating team to turn in a single joint assignment. This system fosters teamwork, builds confidence, and makes the harder problems manageable.

The problems in this book have been classroom-tested according to two different schemes. Problems can be assigned in big chunks, as projects to be worked on three or four times during the term. Alternatively, problems can be assigned one or two at a time on a more regular basis. Both methods work; which works better depends on the backgrounds of instructor and students, and on how thoroughly you want to combine the material from this book with assignments from a standard textbook.

A Word About Versions of *Mathematica*

This book was written using *Mathematica*, Version 3. Many aspects of *Mathematica* changed between Version 2.2 and Version 3.0. *Mathematica*, in addition to its mathematical capabilities, contains many word processing features. In particular, you can format both text and input cells in Version 3.0 Notebooks. In our examples, we have formatted text cells, but not input cells. The reason is simple: we want to make the book accessible to readers who are still using *Mathematica*, Version 2.2, in which formatting of input cells is not possible.

While the *kernel* (which does the actual calculation) has also changed, most of our input cells can be entered in *Mathematica*, Version 2.2 to elicit essentially the same response. Here are some of the kernel changes that you should be aware of if you try to solve the problems in *Mathematica*, Version 2.2:

- The Version 3 command `Cross`, for the cross product of vectors, does not exist in *Mathematica*, Version 2. For purposes of this book, you can replace it by `CrossProduct` in the package `Calculus`VectorAnalysis`.
- The Version 3 integration packages are far more powerful than those in Version 2. In *Mathematica*, Version 2, it is necessary in some cases to load auxiliary packages such as `Calculus`EllipticIntegrate` or `Calculus`DSolve`.
- *Mathematica*, Version 3 has some algebraic simplification commands, such as `ExpToTrig`, which were not available in *Mathematica*, Version 2. In most cases, you can substitute commands such as `ComplexExpand` or options such as `Trig -> True`.
- The algorithm used for `FindMinimum` works differently in *Mathematica*, Version 2 and *Mathematica*, Version 3, and may produce different answers.
- In *Mathematica*, Version 3, but not in *Mathematica*, Version 2, the command `Sum` can sum a large number of finite and infinite series in closed form.
- The default number of steps taken by the numerical differential equations solver, `NDsolve`, has changed in *Mathematica*, Version 3. For some ODEs that can be solved without difficulty in Version 3, you need in Version 2 to increase the value of `MaxSteps`.

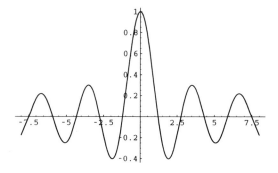

Problem Set A

REVIEW OF ONE-VARIABLE CALCULUS

All problems should be solved in *Mathematica* Notebooks, combining text, graphics, and mathematical computations. All of your explanations should be well-organized and clearly presented in text cells. You should use the options available with the plotting routines to enhance your plots. See the *Sample Notebook Solutions* at the end of the book for models.

1. Graph the following transcendental functions. Use your judgment and some experimentation to find an appropriate range of values of x so that the "main features" of the graph are visible.

   (a) $\sin x$.

   (b) $\tan x$. (Hint: Adjust the `PlotRange` to get a good picture.)

   (c) $\ln x$. (Hint: If your range of values of x includes 0, you may get various error messages. Why? You can bypass the problem by letting x start somewhere like 0.0001.)

   (d) $\sinh x$.

   (e) $\tanh x$.

   (f) e^{-x^2}.

2. Let $f(x) = x^3 - 4x^2 + 1$.

    (a) Graph f.

    (b) Use `Solve` to find (in terms of square and cube roots) the exact values of x where the graph crosses the x-axis. Don't be surprised if the answers are complicated and involve complex numbers.

    (c) Convert the exact x-intercepts (from (b)) to numerical values, and order them as $x_0 < x_1 < x_2$. (The intercepts are all real, even though it may not look that way. Use `Chop` to get rid of the round-off errors.)

    (d) Compute the exact value of the area of the bounded region lying below the graph of f and above the x-axis. (This is the region where $x_0 \leq x \leq x_1$.) Then convert this exact expression to a numerical value and explain in terms of your picture in (a) why the answer is reasonable.

    (e) Determine where f is increasing and where it is decreasing. Use `Solve` to find the exact values of x where f has a relative maximum or relative minimum point.

    (f) Find numerical values of the coordinates of the relative maximum points and/or relative minimum points on the graph.

    (g) Determine where the graph of f is concave upward and where it is concave downward.

3. Consider the equation $x \sin x = 1$, for x a positive number.

    (a) By graphing the function $x \sin x$, find the approximate location of the first five solutions.

    (b) Why is there a solution close to $n\pi$ for every positive integer n? (Hint: For large x, the reciprocal $1/x$ is pretty close to 0. Where is $\sin x = 0$?)

    (c) Use `FindRoot` to refine your approximate solutions from (a) to get numerical values of the true solutions (good to at least several decimal places).

4. Compute the following limits:

    (a) $\lim\limits_{x \to 1} \dfrac{x^2 + 3x - 4}{x - 1}$.

    (b) $\lim\limits_{x \to 0} \dfrac{\sin x}{x}$.

    (c) $\lim\limits_{x \to 0+} x \ln x$.

5. Compute the following derivatives:

   (a) $\dfrac{d}{dx}(x^2 + 5x - 1)^{100}$.

   (b) $\dfrac{d}{dx}\left(\dfrac{x^2 e^x - 1}{x^2 + 2}\right)$.

   (c) $\dfrac{d}{dx}(\sin^5 x \cos^3 x)$.

   (d) $\dfrac{d}{dx}\arctan(e^x)$.

6. Compute the following integrals. Whenever possible, find an **exact** expression. If *Mathematica* cannot compute a definite integral exactly, compute a numerical value using **NIntegrate**.

   (a) $\displaystyle\int (x^2 + 5x - 1)^{10}(2x + 5)\, dx$.

   (b) $\displaystyle\int \arcsin(x)\, dx$.

   (c) $\displaystyle\int_0^1 x^5 (1 - x^2)^{3/2}\, dx$.

   (d) $\displaystyle\int_0^1 x^5 (1 - x)^{3/2}\, dx$.

   (e) $\displaystyle\int_0^\infty x^5 e^{-x^2}\, dx$.

   (f) $\displaystyle\int_0^1 e^{e^x}\, dx$.

   (g) $\displaystyle\int e^x \cos^3 x \sin^2 x\, dx$.

   (h) $\displaystyle\int_0^1 \sin\left(x^3\right)\, dx$.

7. Use the **Series** command to find the Taylor series of the function $\sec x$ around the point $x = 0$, up to and including the term in x^{10}.

8. The alternating harmonic series
$$1 - \frac{1}{2} + \frac{1}{3} - \frac{1}{4} + \frac{1}{5} + \cdots$$
   is known to converge (slowly!!) to $\ln 2$.

   (a) Test this by adding the first 100 terms of the series and comparing with the value of $\ln 2$. Do the same with the first 1000 terms.

(b) The alternating series test says that the error in truncating an alternating series (whose terms decrease steadily in absolute value) is less than the absolute value of the last term included. Check this in the situation of (a). In other words, verify that the difference between $\ln 2$ and the sum of the first 100 terms of the series is less than $\frac{1}{100}$ in absolute value, and that the difference between $\ln 2$ and the sum of the first 1000 terms of the series is less than $\frac{1}{1000}$ in absolute value.

9. The harmonic series

$$1 + \frac{1}{2} + \frac{1}{3} + \frac{1}{4} + \frac{1}{5} + \cdots$$

diverges (slowly!!), by the integral test. This test also implies that the sum of the first n terms of the series is approximately $\ln n$. Test this by adding the first 100 terms of the series and comparing with the value of $\ln 100$, and by adding the first 1000 terms of the series and comparing with the value of $\ln 1000$. Do you see any pattern? If so, test it by replacing 1000 by 10000.

10. The power series

$$\sum_{n=0}^{\infty} (-1)^n \frac{x^{2n}}{(n!)^2}$$

converges for all x to a function $f(x)$.

(a) Let $f_k(x)$ be the sum of the series out to the term $n = k$. Graph $f_k(x)$ for $-8 < x < 8$ with $k = 10, 20, 40$. Superimpose the last two plots. Why is the plot of $f_{40}(x)$ for $-8 < x < 8$ visually indistinguishable from the plot of $f_{20}(x)$ over the same domain?

(b) Apply the $\mathtt{Sum}$ command in *Mathematica* to the infinite series. You should find that *Mathematica* recognizes $f(x)$. (However, it may not be a function you saw in your one-variable calculus class.) Plot $f(x)$ for $-8 < x < 8$ and compare with your plot of $f_{40}(x)$ as a means of checking your answer to (a).

11. The $\mathtt{PolarPlot}$ command is contained in the $\mathtt{Graphics\,\textquotesingle Graphics\,\textquotesingle}$ package. In order to use the command, you must first load the package by typing $\mathtt{<<Graphics\,\textquotesingle Graphics\,\textquotesingle}$. Load the package, and then graph the following equations in polar coordinates:

(a) $r = \sin\theta$.

(b) $r = \sin(6\theta)$.

(c) $r = 4\sin\theta - 2$.

(d) $r^2 = \sin 2\theta$.

Glossary of Some Useful *Mathematica* Objects

Commands

`Chop` Chops off small round-off errors.

`D` Differentiates an expression. The symbol ∂ from the standard palette has the same effect. We can also use a prime. For example, `f'[t]` is the derivative of `f[t]`.

`Expand` Expands a complex algebraic expression.

`Factor` Factors an algebraic expression.

`FindRoot` Solves equations numerically, using a variant of Newton's method.

`Integrate` Integrates an expression. The symbol $\int$ from the standard palette has the same effect.

`Limit` Evaluates the limit of an expression.

`N` Evaluates an expression numerically. You can specify the desired accuracy.

`NIntegrate` Numerically approximates a definite integral (using something akin to Simpson's rule, but more complicated).

`Plot` Graphs a function.

`PolarPlot` Plots in polar coordinates.

`Series` Computes the Taylor series of a function around a specified point, out to a specified order.

`Show` Displays several graphics together.

`Simplify` Simplifies complex algebraic expressions.

`Solve` Finds the solutions of an algebraic equation (in terms of radicals, etc.).

`Sort` Sorts a list, putting numbers in numerical order.

`Sum` Sums a series.

`Table` Creates a list.

`//` Sends the output of one command through another. For example, you can add `// N` to the end of an input line to ask for a numerical value; or you can add `// Simplify` to the end to ask for a simplified expression.

/. Replacement operator; e.g., the input

```
x^3 + x /.  x -> 1
```

tells *Mathematica* to replace all occurrences of the variable **x** by the value **1** in the expression **x^3 + x**, yielding the answer 2.

Options and Directives

AspectRatio Height-to-width ratio of a plot. The default is the reciprocal of the golden ratio, about 0.6.

Automatic When used as a value for **AspectRatio**, forces the same scale on both axes.

Direction Option for **Limit**, making it possible to ask for one-sided limits.

PlotPoints The number of points where a function is sampled in plotting. The default value for **Plot** is 25.

PlotRange Specifies the range of values of the dependent variable in a plot.

Built-in Functions

Abs Absolute value function.

ArcSin Inverse sine function.

ArcCos Inverse cosine function.

ArcTan Inverse tangent function.

Cos Cosine function.

Cosh Hyperbolic cosine function, $\cosh(x) = (e^x + e^{-x})/2$.

Exp Exponential function; **Exp[x]** is the same as **E^x**.

Factorial The factorial of a number. Can also be denoted by an exclamation point **!**.

Log The *natural* logarithm, or logarithm to the base e. **Log[10, x]** is $\log_{10} x$.

Sec Secant function.

Sin Sine function.

Sinh Hyperbolic sine function, $\sinh(x) = (e^x - e^{-x})/2$.

Sqrt Square root function. The symbol $\sqrt{}$ from the standard palette has the same effect.

`Tan` Tangent function.

`Tanh` Hyperbolic tangent function, $\tanh(x) = \sinh(x)/\cosh(x)$.

Built-in Constants

`E` The number $e = 2.718281828\ldots$.

`I` The imaginary unit $i = \sqrt{-1}$. Even in problems that don't seem to involve complex numbers, *Mathematica* sometimes will find complex roots to equations or will express integrals in terms of complex-valued functions.

`Infinity` Self-explanatory. *Mathematica* recognizes this as a valid limit in definite integrals, sums, etc. The symbol ∞ from the standard palette has the same meaning.

`Pi` The number π. The symbol π from the standard palette has the same meaning.

Packages

`Graphics`Graphics`` Needed for `PolarPlot`.

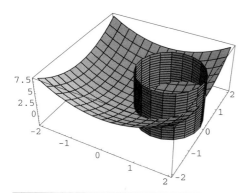

Chapter 2

VECTORS AND GRAPHICS

We start this chapter by explaining how to use vectors in *Mathematica*, with an emphasis on practical operations on vectors in the plane and in space. We discuss the standard vector operations, and give several applications to the computations of geometric quantities such as distances, angles, areas, and volumes. The bulk of the chapter is devoted to instructions for graphing curves and surfaces.

Vectors

In *Mathematica*, vectors are represented as lists of numbers or variables. You write a list in *Mathematica* as a sequence of entries encased in braces. Thus, you would enter **v = {3, 2, 1}** at the prompt to tell *Mathematica* to treat **v** as a vector with x, y, z coordinates equal to 3, 2, and 1, respectively.

You can perform the usual vector space operations in *Mathematica*: vector space addition, scalar multiplication, and the dot product. Here are some examples:

```
In[1]:=  a := {1, 2, 3};
         b := {-5, -3, -1};
         c := {3, 0, -2};

In[2]:=  a + b

Out[2]=  {-4, -1, 2}
```

```
In[3]:=  5c
```

Out[3]= $\{15, 0, -10\}$

```
In[4]:=  a.b
```

Out[4]= -14

The period key is used to compute the dot product. (Alternatively, you can write `Dot[a, b]`.) As usual, you can use the dot product to compute lengths of vectors (also known as vector *norms*).

```
In[5]:=  lengthofa = N[Sqrt[a.a]]
```

Out[5]= 3.74166

Here is a function that automates the numerical computation of vector norms.

```
In[6]:=  norm[v_] := N[Sqrt[v.v]]
```

```
In[7]:=  norm[a]
         norm[b]
         norm[c]
```

Out[7]= 3.74166

Out[8]= 5.91608

Out[9]= 3.60555

Our attention throughout this book will be directed to vectors in the plane and vectors in (three-dimensional) space. Vectors in the plane have two components; a typical example in *Mathematica* is `{x, y}`. Vectors in space have three components, like `{x, y, z}`. The following principle will recur: *Vectors with different numbers of components don't mix.* As you will see, certain *Mathematica* commands will only work with two-component vectors; others will only work with three-component vectors. To convert a vector in the plane into a vector in space, you can add a zero to the end.

```
In[10]:= pl = {x, y}
```

Out[10]= $\{x, y\}$

```
In[11]:= sl = Join[pl, {0}]
```

Out[11]= $\{x, y, 0\}$

To project a vector in space into a vector in the x-y plane, you simply drop the final component.

In[12]:= **s2 = {x, y, z}**

Out[12]= $\{x, y, z\}$

In[13]:= **p2 = Drop[s2, -1]**

Out[13]= $\{x, y\}$

The first place we notice the difference is with the cross product, which only works on a pair of three-dimensional vectors. To compute the cross product in *Mathematica*, you have two choices. With *Mathematica* 3.0, you can use the `Cross` command. In all versions of *Mathematica*, you can use the `CrossProduct` command from the `Calculus`VectorAnalysis`` package. To load this package, type the following:

In[14]:= **<<Calculus`VectorAnalysis`**

Then to compute a cross product, type

In[15]:= **CrossProduct[a, b]**

Out[15]= $\{7, -14, 7\}$

Applications of Vectors

Since *Mathematica* allows you to perform all the standard operations on vectors, it is a simple matter to compute lengths of vectors, angles between vectors, distances between points and planes or between points and lines, areas of parallelograms, and volumes of parallelepipeds, or any of the other objects encountered in the introductory vector theory portion of a typical text. Here are some examples.

THE ANGLE BETWEEN TWO VECTORS. The fundamental identity involving the dot product is

$$\mathbf{a} \cdot \mathbf{b} = \|\mathbf{a}\| \, \|\mathbf{b}\| \cos(\varphi),$$

where φ is the angle between the two vectors. So, we can find the angle between **a** and **b** in *Mathematica* by typing

In[16]:= **phi = ArcCos[a.b/(norm[a]*norm[b])]**

Out[16]= 2.25552

We can convert this value from radians to degrees by typing

In[17]:= **phi*180/Pi**

Out[17]= 129.232

THE VOLUME OF A PARALLELEPIPED. The volume of the parallelepiped spanned by the vectors **a**, **b**, and **c** is computed using the formula $V = |\mathbf{a} \cdot (\mathbf{b} \times \mathbf{c})|$. In *Mathematica*, it is easy to enter this formula.

In[18]:= **Abs[a.CrossProduct[b, c]]**

Out[18]= 7

THE AREA OF A PARALLELOGRAM. The volume formula becomes an area formula if you take one of the vectors to be a unit vector perpendicular to the plane spanned by the other two vectors. In particular, the area of the parallelogram spanned by the vectors **a** and **c** is just given by $\|a \times c\|$, which you compute in *Mathematica* by typing

In[19]:= **norm[CrossProduct[a, c]]**

Out[19]= 13.1529

In a similar fashion, you can use the standard mathematical formulas in *Mathematica* to:

(i) project a vector onto a line or a plane;

(ii) compute the distance from a point to a plane;

(iii) compute the distance from a point to a line; and

(iv) check that lines and planes are parallel or perpendicular.

Parametric Curves

We have assumed that you already know how to use *Mathematica*'s **Plot** command to graph plane curves of the form $y = f(x)$. In this section, we will explain how to graph curves defined by parametric equations, both in the plane and in space. As with many of *Mathematica*'s graphing commands, there is a pair of related commands that you can use to accomplish this goal. The **ParametricPlot** command graphs parametric curves in the plane; **ParametricPlot3D** graphs curves in space.

Let's start by looking at the plane curve defined parametrically by the equations

$$x = e^{-t/10}(1 + \cos(t)), \quad y = e^{-t/10}\sin(t), \quad t \in \mathbf{R}.$$

These equations mean that t is a parameter (ranging through the real numbers), and that associated to each value of t is a pair (x, y) of values defined by the given formulas. As t varies, the points $(x(t), y(t))$ trace out a curve in the plane. We define this parametric curve in *Mathematica* by creating a vector whose entries are the expressions defining x and y.

In[20]:= **curve = {Exp[-t/10]*(1 + Cos[t]),**
          Exp[-t/10]*Sin[t]}

Out[20]= $\{E^{-t/10}(1 + \text{Cos}[t]), E^{-t/10}\text{Sin}[t]\}$

Here is a graph of a small portion of the curve.

In[21]:= **ParametricPlot[Evaluate[curve], {t, 0, 2Pi}];**

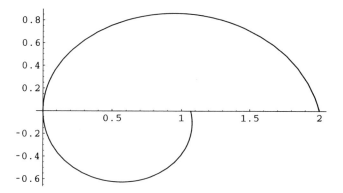

You should almost always use the **Evaluate** command on the first argument to the **ParametricPlot** command; otherwise, *Mathematica* will generate a string of warning messages about uncompiled functions. (See Question 9 in Chapter 10, *Mathematica Tips*.)

Here is a preliminary attempt to draw a larger portion of the curve.

```
In[22]:= ParametricPlot[Evaluate[curve], {t, 0, 20Pi}];
```

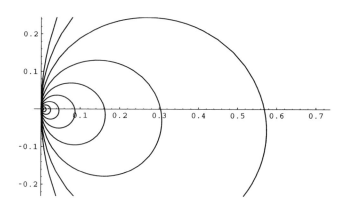

In this graph, *Mathematica* has cut off part of the curve in order to show more detail in other parts. To force *Mathematica* to show all of the curve, you can use the **PlotRange** option.

```
In[23]:= ParametricPlot[Evaluate[curve], {t, 0, 20Pi},
            PlotRange -> All];
```

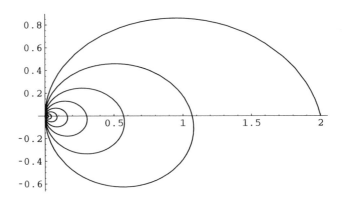

By typing `PlotRange -> {{a, b}, {c, d}}`, you can set explicit ranges on both the x- and y-axes in the plot.

Now let's look at a space curve defined by a set of parametric equations. As our example, we will take Viviani's curve, which is defined as the intersection of the sphere $x^2 + y^2 + z^2 = 4$ with the cylinder $(x - 1)^2 + y^2 = 1$. The projection of Viviani's curve into the x-y plane is defined by the same equation that defines the cylinder; thus, it is a circle that has been shifted away from the origin. We can parametrize this circle in the plane by taking

$$x = 1 + \cos(t), \quad y = \sin(t), \quad -\pi \le t \le \pi.$$

Using this parametrization to solve the sphere's equation for z, we find that

$$z = \pm\sqrt{4 - x^2 - y^2}$$
$$= \pm\sqrt{4 - (1 + \cos(t))^2 - \sin^2(t)}$$
$$= \pm\sqrt{2 - 2\cos(t)}$$
$$= 2\sin\left(\frac{t}{2}\right).$$

Now we will define Viviani's curve in *Mathematica*. After that, we will use `ParametricPlot3D` to graph the part of the curve in the first octant.

```
In[24]:= viviani = {1 + Cos[t], Sin[t], 2Sin[t/2]}
```

$$\text{Out[24]= } \left\{1 + \text{Cos[t], Sin[t], 2Sin}\left[\frac{t}{2}\right]\right\}$$

```
In[25]:= ParametricPlot3D[Evaluate[viviani], {t, 0, Pi}];
```

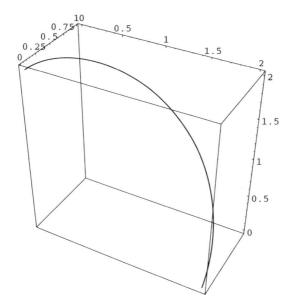

That picture isn't terribly illuminating; let's see if we can improve it. The first thing we are going to do is to get rid of the box and the tick marks that, by default, surround all three-dimensional graphics in *Mathematica*. We'll do that by setting **Boxed -> False** and **Ticks -> None**. It would also be nice to show the curve inside the sphere and cylinder that are used to define it. The arcs obtained by intersecting the sphere with the coordinate planes are parametrized by

```
In[26]:= arcs = {{0, 2Cos[t/2], 2Sin[t/2]},
        {2Cos[t/2], 0, 2Sin[t/2]},
        {2Cos[t/2], 2Sin[t/2], 0}}
```

$$\text{Out[26]= } \left\{\left\{0, 2\text{Cos}\left[\frac{t}{2}\right], 2\text{Sin}\left[\frac{t}{2}\right]\right\}, \left\{2\text{Cos}\left[\frac{t}{2}\right], 0, 2\text{Sin}\left[\frac{t}{2}\right]\right\},\right.$$
$$\left.\left\{2\text{Cos}\left[\frac{t}{2}\right], 2\text{Sin}\left[\frac{t}{2}\right], 0\right\}\right\}$$

The projected circle in the x-y plane is parametrized as described above.

```
In[27]:= circ = {1 + Cos[t], Sin[t], 0}
```

$$\text{Out[27]= } \{1 + \text{Cos}[t], \text{Sin}[t], 0\}$$

Now we can collect all the curves we'd like to plot into a single list.

```
In[28]:= total = Join[{viviani}, {circ}, arcs];
```

(We have terminated the input line with a semicolon to prevent *Mathematica* from printing output that we really don't need to see.) Here's our second attempt to make an informative plot.

```
In[29]:= ParametricPlot3D[Evaluate[total], {t, 0, Pi},
          Boxed -> False, Ticks -> None];
```

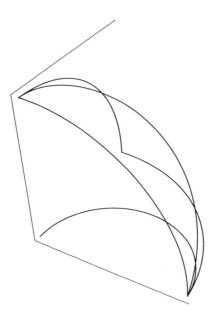

That still isn't very good; however, a few minor changes will improve it signifi-
cantly. The main problem is the viewpoint from which *Mathematica* has chosen to
show us the graph. The default viewpoint lies at the point $(1.3, -2.4, 2)$, where the
units are multiples of the length of each displayed axis. This viewpoint is chosen
generically, to make it unlikely that significant features of a random graph will be
obscured. It has the definite disadvantage, however, that it changes the apparent
directions of the x- and y-axes when compared with most mathematical textbooks.
Choosing all positive values for the viewpoint will put your viewpoint into the first
octant with the axes proceeding in the usual directions. In this case, we'll make
the change `ViewPoint -> {10, 3, 1}` to get a better look at the graph. (In
many cases, `{1, 1, 1}` is a good choice; you may need to experiment to find the
best viewpoint.) Second, we'll tell *Mathematica* to draw the axes at the back of
the graph. Finally, we'll set the `PlotRange` explicitly to the actual range of values
needed on each axis.

```
In[30]:= ParametricPlot3D[Evaluate[total], {t, 0, Pi},
          Boxed -> False, Ticks -> None,
          ViewPoint -> {10, 3, 1},
          AxesEdge -> {{-1, -1}, {-1, -1}, {-1, -1}},
          PlotRange ->{{0, 2}, {0, 2}, {0, 2}}];
```

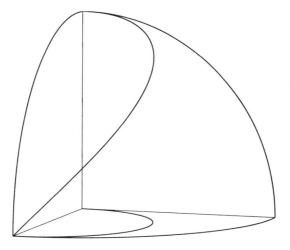

Graphing Surfaces

Our next goal is to learn how to graph surfaces that are defined by a single equation $z = f(x, y)$. The command to do that is **Plot3D**, the three-dimensional analog of **Plot**. Let's look at a simple example. We'll plot the function $f(x, y) = 1 - (x^2 + y^2)$ on the rectangle $-1 \leq x \leq 1, -1 \leq y \leq 1$.

```
In[31]:= surf = 1 - (x^2 + y^2)
```

$$\text{Out[31]= } 1 - x^2 - y^2$$

```
In[32]:= pic1 = Plot3D[surf, {x, -1, 1}, {y, -1, 1}];
```

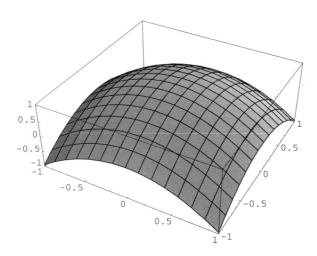

Unlike its two-dimensional analog, the `Plot3D` command will not allow you to plot more than one surface at the same time. You need to generate the pictures separately and then combine them with the `Show` command. For example, let's draw a second graph on the same rectangle.

In[33]:= **pic2 = Plot3D[Sin[6*x*y], {x, -1, 1}, {y, -1, 1}];**

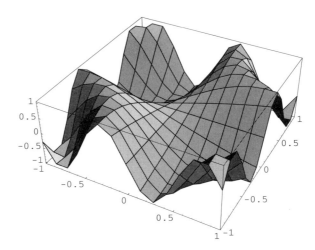

Now we can combine the two graphs. In the process, we'll reset some of the options.

In[34]:= **Show[{pic1, pic2}, Ticks -> None,**
         AxesLabel -> ["x", "y", "z"]];

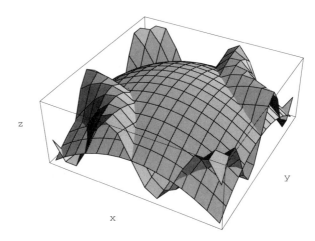

Note here that *Mathematica* treats surfaces as being opaque—when one surface lies in front of another, the one in back is obscured by the one in front.

The `Plot3D` command requires that the region in the x-y plane, over which the surface is plotted, must be a rectangle. Later on, we will see that it is helpful to be able to visualize portions of surfaces that lie over curved regions in the x-y plane. This will be especially important when we study multiple integrals. In the meantime, here is a simple scheme to implement such a drawing. Define an auxiliary function $g(x, y)$ as follows:

In[35]:= `g[x_, y_] = If[x^2 + y^2 < 1, 1, 0]`

Out[35]= $If[x^2 + y^2 < 1, 1, 0]$

You can consult the online help for more information on the `If` command. For now, suffice it to say that in this definition, we have set $g(x, y)$ equal to 1 if the condition $x^2 + y^2 < 1$ is satisfied, and equal to 0 otherwise.

In[36]:= `Plot3D[surf*g[x, y], {x, -1, 1}, {y, -1, 1}];`

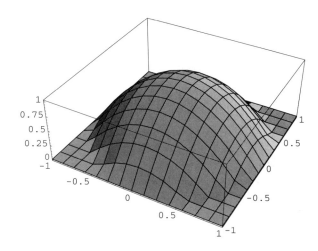

Parametric Surfaces

The final topic is the plotting of surfaces that are defined by parametric equations. A parametric curve, being one-dimensional, depends on a single parameter. A parametric surface, being two-dimensional, requires us to use two parameters. Let's denote the parameters by u and v. For each parameter pair (u, v), we need to specify an associated point $(x(u, v), y(u, v), z(u, v))$ in space. As the parameter point (u, v) varies over its domain (which is some region in the plane), the associated point will trace out a surface in three-dimensional space. As an example, we will consider the surface whose parametric equations are

$$x = u^3, \quad y = v^3, \quad z = uv.$$

We often express this notion mathematically by writing

$$(x, y, z) = (u^3, v^3, uv).$$

In *Mathematica*, we can define

```
In[37]:= parasurf = {u^3, v^3, u*v}
```

Out[37]= $\{u^3, v^3, u\,v\}$

```
In[38]:= ParametricPlot3D[Evaluate[parasurf], {u, -1, 1},
        {v, -1, 1}, ViewPoint -> {1, 1, 1}, Boxed -> False,
        Axes -> False];
```

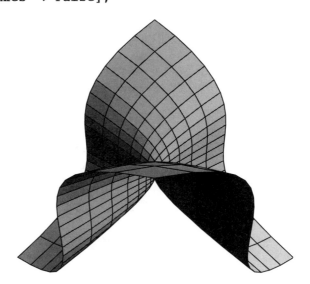

The standard methods for improving two-dimensional plots can also be employed for three-dimensional plots. For example, if your surfaces or curves look jagged or ill-defined, you can have *Mathematica* sample more points by increasing the value of `PlotPoints`. Depending on your first picture, you may be able to estimate that the range you supplied (for the parameter or the base rectangle) should be adjusted. It is a simple matter to edit the input cell and then reinvoke the command. Your first picture will be overwritten with a second picture. Labeling the axes, labeling the picture, changing the `AspectRatio` or `ViewPoint`—these, and other techniques that you will learn to use as you produce more graphs, can greatly improve the quality of your plots.

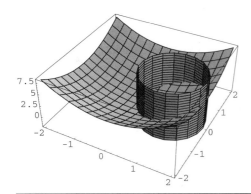

Problem Set B

VECTORS AND GRAPHICS

1. Find the distance between the two points $P = (0, 4.516, -5.298)$ and $Q = (-3.33, 0.234, 7.8)$.

2. Show that the point $P = (2, 0, 3)$ is equidistant from the two points $Q1 = (0.12, -1, 5.55)$ and $Q2 = (3.88, 1, 0.45)$.

3. Suppose that the points $P1 = (-2, -3, 5)$ and $P2 = (-6, 3, 1)$ are the endpoints of a diagonal in a sphere. Find the equation of the sphere. Then compute the coordinates of any points on the line $z = 10y = -x$ which also lie on the surface of the sphere.

4. Let $\mathbf{a} = (2.9999, 400001, -6)$ and $\mathbf{b} = (0, -3.8765, 592320)$. Find $\mathbf{a} + \mathbf{b}$, $\|\mathbf{b}\|$ and $7\mathbf{a}$. Explain why $\|\mathbf{b}\|$ apparently equals the z-coordinate of $\mathbf{b}$, even though the y-coordinate is not zero.

5. Compute the angle (in degrees) that the vector $\mathbf{a} = -24.56\mathbf{i} + 44.689\mathbf{j}$ makes with the x-axis, measured counterclockwise from the x-axis.

6. Two tugboats are pulling a cruise ship. Tugboat 1 exerts a force of 1000 pounds on the ship and pulls in the direction 30 degrees north of due east. Tugboat 2 pulls in the direction 45 degrees south of due east. What force must Tugboat 2 exert to keep the ship moving due east? (If you want to draw a picture, the

package `Graphics'Arrow'` is convenient for this purpose.)

7. Consider the vectors
$$\mathbf{a} = (9, -3, 0.25),$$
$$\mathbf{b} = (-3, -4, 60),$$
$$\mathbf{c} = (-20.4, -6.2, 155.65).$$

(a) Verify that **a** and **b** are perpendicular.

(b) The vector **c** lies in the same plane as **a** and **b**. Resolve the vector **c** into its **a** and **b** components. Check your answer.

8. Find the angle (in degrees) between each of the following pairs of vectors:

(a) $\mathbf{a} = (2.467, -4.196, 0.433)$ and $\mathbf{b} = (-10.43, 9.344, 0)$; and

(b) $\mathbf{a} = (-3.54, -10.79, 0.991)$ and $\mathbf{b} = (-1.398, 0, 6.443)$.

9. The following four vectors lie in a plane in 3-space. Do their endpoints determine a parallelogram? a rhombus? a square?
$$\mathbf{a} = (1, 1, 1), \qquad \mathbf{b} = (2, 3, 3), \qquad \mathbf{c} = (4, 2, 5), \qquad \mathbf{d} = (3, 0, 3).$$

10. In each of the following cases, find the cross product of the vectors a and b, and then use it to find the angle between the two vectors.

(a) $\mathbf{a} = (-4.275, -2.549, 9.333), \quad \mathbf{b} = (6.302, -2.043, 0.444)$.

(b) $\mathbf{a} = (77, 88, 99), \quad \mathbf{b} = (22, 44, 66)$.

11. Prove the identity
$$\|\mathbf{a} \times \mathbf{b}\|^2 = \|\mathbf{a}\|^2 \|\mathbf{b}\|^2 - (\mathbf{a} \cdot \mathbf{b})^2$$
by establishing letter (i.e., variable) coordinates in a and b and evaluating both sides of the identity using *Mathematica*.

12. In this problem, we study the volumes of parallelepipeds.

(a) Find the volume of the parallelepiped determined by the three vectors:
$$\mathbf{a} = (8324, 5789, 2098),$$
$$\mathbf{b} = (9265, -246, 8034),$$
$$\mathbf{c} = (4321, -765, 7903).$$

(b) Now consider all parallelepipeds whose base is determined by the vectors $\mathbf{a} = (2, 0, -1)$ and $\mathbf{b} = (0, 2, -1)$, and whose height is variable $\mathbf{c} = (x, y, z)$. Assume that x, y, and z are positive and $\|\mathbf{c}\| = 1$. Use the triple product to compute a formula for the volume of the parallelepiped involving x, y, and z. Compute the maximum value of that volume in terms of x and y as follows. It is clear from the following formula

$$\mathbf{c} \cdot (\mathbf{a} \times \mathbf{b}) = \|\mathbf{c}\| \, \|\mathbf{a} \times \mathbf{b}\| \cos\theta,$$

where θ is the angle between $\mathbf{c}$ and the line perpendicular to the plane determined by $\mathbf{a}$ and $\mathbf{b}$, that the maximum occurs when $\mathbf{c}$ is perpendicular to both $\mathbf{a}$ and $\mathbf{b}$. Use the dot product to determine the vector $\mathbf{c}$ yielding the maximum value. (We will see in Chapter 7, *Optimization in Several Variables*, how to solve multivariable max-min problems.)

13. Find parametric equations for each of the following lines, then graph the line using `ParametricPlot3D`.

 (a) The line containing the points $(5.2, -4.11, 9)$ and $(0.3, 6.33, -2.34)$.

 (b) The line passing through the point $(4, 0.35, -3.72)$ and parallel to the vector $\mathbf{v} = (4.66, -2.1, -3.51)$.

14. Find the distance from the point $(4.3, 5.4, 6.5)$ to the line whose parametric equations are $x = -1 + t$, $y = -2 + 2t$, $z = -3 + 3t$.

15. Draw the cylinder whose points lie at a distance 1 from the line $x = t$, $y = 10t$, $z = 0$. (Hint: To use `ParametricPlot3D`, choose two unit vectors perpendicular to the line and use them and a vector along the line to parametrize the cylinder.)

16. For each of the following, find the equation of the plane and graph it:

 (a) The plane containing the point $P_0 = (3.4, -2.6, 5)$ and having normal vector $\mathbf{n} = (-3.22, 1.2, 0.3)$; and

 (b) The plane containing the two lines

$$x = 1 + t, \qquad y = 2 + t, \qquad z = 1 + 2t,$$
$$x = 2t, \qquad y = 1 + t, \qquad z = -1 - t.$$

17. Find the distance from the point $P = (100, 201, 349)$ to the plane $-213x - 438y + 301z = 500$.

18. Find parametric equations for the line formed by the intersection of the following two planes:

$$2x - 3y + z = 10,$$
$$-5x - 2y + 3z = 15.$$

Graph the two planes and the line of intersection on the same plot.

19. Consider the vector-valued functions

$$F(t) = (e^t, \sqrt{1+t}, \ln(1+t^2)),$$
$$G(t) = (\sin(t), \sec(t+1), (t-1)/(t+1)).$$

Compute the functions $\mathbf{F} + \mathbf{G}$, $\mathbf{F} \cdot \mathbf{G}$, and $\mathbf{F} \times \mathbf{G}$.

20. Plot the following curves. In each case indicate the direction of motion. You will have to be careful when selecting the time interval on which to display the curve in order to get a meaningful picture. You may find the `PlotRange` option useful in your plots.

    (a) $\mathbf{F}(t) = (\cos t, \sin t, t/2)$.

    (b) $\mathbf{F}(t) = (e^{-t} \sin t, e^{-t} \cos t, 1)$.

    (c) $\mathbf{F}(t) = (t, t^2, t^3)$.

21. Graph the cycloid

$$\mathbf{r}(t) = (2(t - \sin t), 2(1 - \cos t))$$

and the trochoid

$$\mathbf{s}(t) = (2t - \sin t, 2 - \cos t)$$

together on the interval $[0, 4\pi]$. Find the coordinates of the four points of intersection. (Hint: Solve the equation $\mathbf{r}(t) = \mathbf{s}(u)$. Note the different independent variables for $\mathbf{r}$ and $\mathbf{s}$—the points of intersection need not correspond to the same "time" on each curve. Also, since the coordinate functions are transcendental, you may need to use `FindRoot` rather than `Solve`.)

22. Here's a problem to practice simultaneous plotting of curves and surfaces, as well as finding intersection points.

    (a) Plot the two curves $2x^2 + 20y = -1$ and $y = x^4 - x^2$ on the same graph. Find the coordinates of all points of intersection.

    (b) Plot the two surfaces $x^2 + y^2 + z^2 = 16$ and $z = 4x^2 + y^2$ and superimpose the plots. (The first surface is a sphere, the second is an elliptic paraboloid. The top half of the sphere will suffice here.) Use `ImplicitPlot` to plot the projection into the x-y plane of the curve of intersection of the two surfaces. (You should load the package `<<Graphics'ImplicitPlot'` and look up its syntax, either in the Help Browser or by typing `?ImplicitPlot`.)

23. This problem is about intersecting surfaces and curves. For helpful models, see the discussions of Viviani's curve, of "Graphing Surfaces," and of "Parametric Surfaces" in Chapter 2, *Vectors and Graphics*. You might wish to adjust the `ViewPoint` in each of your three-dimensional plots.

(a) Draw three-dimensional plots of the paraboloid $z = x^2 + y^2$ and of the cylinder $(x - 1)^2 + y^2 = 1$. Since the cylinder is not given by an equation of the form $z = f(x, y)$, you will need to plot it parametrically. You can use the parametrization

$$(1 + \cos t, \sin t, z)$$

with parameters t and z. Superimpose the two three-dimensional plots to see the curve where the surfaces intersect. Find a parametrization of the curve of intersection, and then draw a nice three-dimensional plot of this curve.

(b) Do the same for the paraboloid $z = x^2 + y^2$ and the upper hemisphere $z = \sqrt{1 - (x - 1)^2 - y^2}$. This time the equation of the curve of intersection is a bit complicated in rectangular coordinates. It looks better if you project the curve into the x-y plane and convert to polar coordinates. Apply `Solve` to the polar equation of the projection to find $r(\theta)$, the formula for r in terms of θ. You can then parametrize the curve by

$$(r(\theta) \cos \theta, \ r(\theta) \sin \theta, \ r(\theta)^2),$$

with θ varying.

Glossary of Some Useful *Mathematica* Objects

Commands

`Clear` Clears the definition of a variable.

`Cross` Cross product command in *Mathematica* 3.0.

`CrossProduct` Cross product command in the `Calculus 'VectorAnalysis'` package.

`Dot` Dot product. Also written with a period, as in `A.B`.

`Evaluate` Forces evaluation of an expression; needed in some plotting commands.

`FindRoot` Finds numerical solutions of an equation.

`First` First element of a list.

`If` Used for if/then/else conditional statements.

`ImplicitPlot` Plots an implicit equation in the plane.

`Last` Last element of a list.

`N` Evaluates an expression numerically.

`NSolve` Same as `Solve` followed by `N`.

`ParametricPlot` Plots curves in the plane defined by parametric equations.

`ParametricPlot3D` Plots curves or surfaces in space defined by parametric equations.

`Plot3D` Plots the graph of a function of two variables.

`Show` Displays several graphics together.

`Solve` Finds exact solutions of an equation.

`Simplify` Simplifies complex expressions.

`Sum` Computes finite and infinite sums.

`Table` Creates a list; useful for producing vectors.

Options and Directives

`AspectRatio` Height-to-width ratio of a plot.

`AxesEdge` In three-dimensional graphics, specifies on what edges of the bounding box the axes should be drawn.

`Boxed` Specifies whether a three-dimensional graphic is encased in a box. Default is `True`.

`DisplayFunction` Controls how a graph is displayed. Setting `DisplayFunction -> Identity` suppresses the graph; you can then set `DisplayFunction -> $DisplayFunction` in `Show` to combine graphs.

`PlotPoints` The number of points where a function is sampled in plotting. The default value for `Plot` is 25.

`PlotRange` Specifies the range of values of the dependent variable in a plot.

`Ticks` Controls the placement of tick marks along axes in a graph.

`ViewPoint` Specifies a point from which to view three-dimensional graphics.

Packages

`Calculus`VectorAnalysis`` Needed for `CrossProduct`.

`Graphics`Arrow`` Convenient for drawing two-dimensional vectors as arrows.

`Graphics`ImplicitPlot`` Needed for `ImplicitPlot`.

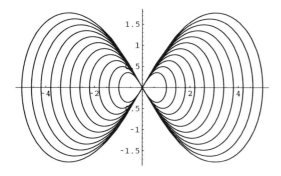

Chapter 3

GEOMETRY OF CURVES

Curves are the most basic geometric objects. In this chapter, we will study curves in the plane and curves in three-dimensional space. To each curve, we can attach certain natural geometric *invariants*; that is, quantities that express physical properties of the curve. These invariants include:

- arclength;
- the number of singularities (such as cusps);
- curvature, which measures how much the curve bends; and
- torsion, which measures how much the curve twists.

We will use calculus to define each invariant in terms of derivatives and integrals. Finally, we will show that the geometric invariants characterize the curve. In other words, distinct curves cannot have identical invariants, unless they are congruent.

Parametric Curves

A curve is the image of a continuous (and usually differentiable) function $\mathbf{r} : I \to \mathbf{R}^n$, where I is an interval. In this book, n is either 2 or 3. We shall be ambiguous about the nature of the interval I, allowing for the possibility of it being open, closed, bounded, or unbounded. Here are some examples of curves.

- The right circular helix of radius 1:

$$\mathbf{r}(t) = (\cos t, \sin t, t), \quad t \in \mathbf{R}.$$

- A three-dimensional astroid:

$$\mathbf{r}(t) = (\cos^3 t, \sin^3 t, \cos 2t), \quad 0 \le t \le 2\pi.$$

- The cycloid is the curve traced out by a point on the circumference of a wheel as it rolls in a straight line at constant speed without slipping. The parametric equations are:

$$\mathbf{r}(t) = (t - \sin t, 1 - \cos t), \quad t \in \mathbf{R}.$$

The cycloid is a plane curve; the helix and astroid are space curves. Throughout the first part of this chapter, we will use the helix as our illustrative example. We will return to the other examples later in the chapter.

We begin by defining the helix in *Mathematica* and then graphing it.

```
In[1]:= helix = {Cos[t], Sin[t], t}
```

Out[1]= $\{\text{Cos}[t], \text{Sin}[t], t\}$

```
In[2]:= helPlot = ParametricPlot3D[Evaluate[helix], {t,
        -2Pi, 2Pi}, ViewPoint -> {1, 1, 1}, Ticks -> None];
```

The function $\mathbf{r}(t)$, $t \in I$, is called a *parametrization* of the curve; the curve itself is the physical set of points (in $\mathbf{R}^2$ or $\mathbf{R}^3$) traced out by the function $\mathbf{r}$ as t varies in the interval I. Every curve has many different possible parametrizations. All of the geometric invariants we associate to a curve will be computed analytically in terms of a parametrization. Clearly, if there is to be any geometric validity for an invariant, the quantity should be independent of the choice of parametrization.

If we think of a particle traversing a curve according to the prescription $\mathbf{r}(t)$, then it is natural to call $\mathbf{r}(t)$ the position vector. Continuing with the physical analogy, we call $\mathbf{v}(t) = \mathbf{r}'(t)$ the velocity vector and $\mathbf{a}(t) = \mathbf{r}''(t)$ the acceleration vector—provided these derivatives exist. The speed $v(t)$ of the particle is defined as the magnitude of the velocity vector, $v(t) = \|\mathbf{v}(t)\|$.

As an example, we can compute the velocity, speed, and acceleration of the helix. We start with a command that we will use repeatedly in the future to compute the length of vectors.

```
In[3]:= vectorLength[x_] := Sqrt[Simplify[x.x]]
```

```
In[4]:= velhelix = D[helix, t]
```
$$\text{Out[4]=} \quad \{-\mathrm{Sin}[t], \mathrm{Cos}[t], 1\}$$

```
In[5]:= speedhelix = vectorLength[velhelix]
```
$$\text{Out[5]=} \quad \sqrt{2}$$

```
In[6]:= acchelix = D[velhelix, t]
```
$$\text{Out[6]=} \quad \{-\mathrm{Cos}[t], -\mathrm{Sin}[t], 0\}$$

We say that a curve is *smooth* if its velocity vector is a continuous function and its speed never vanishes. We can see from the formulas just computed that the velocity vector of the helix only involves continuous functions, and that the speed is the nonzero constant $\sqrt{2}$. So, the helix is a smooth curve. More generally, we say that a curve is *piecewise smooth* if in any bounded subinterval of I, the curve is smooth except for finitely many points at which one-sided derivatives of $\mathbf{r}$ are postulated to exist. In the following discussion we will generally assume our curves are smooth, though piecewise smooth curves with a few singularities will also be considered.

Now suppose we have two parametrizations of the same non-self-intersecting curve:

$$\mathbf{F}(t),\ t \in [a, b] \quad \text{and} \quad \mathbf{G}(u),\ u \in [c, d].$$

Then $\mathbf{F}$ and $\mathbf{G}$ are one-to-one functions and for every $t \in [a, b]$, there is a unique $u \in [c, d]$ such that $\mathbf{F}(t) = \mathbf{G}(u)$, and vice versa. Since the two parametrized curves must have the same endpoints, either $\mathbf{F}(a) = \mathbf{G}(c)$ and $\mathbf{F}(b) = \mathbf{G}(d)$, or else $\mathbf{F}(a) = \mathbf{G}(d)$ and $\mathbf{F}(b) = \mathbf{G}(c)$. We assume the former, which means simply that $\mathbf{F}$ and $\mathbf{G}$ trace out the points of the curve *in the same order*, or in other words, they have the *same orientation*. Hence, there is a uniquely defined, monotonically increasing function $u = \varphi(t)$, $\varphi(a) = c$, $\varphi(b) = d$ so that $\mathbf{G}(\varphi(t)) = \mathbf{F}(t)$, $t \in [a, b]$. Moreover, if the parametrizations are smooth, then φ is differentiable, φ' is continuous and never vanishes. The latter follows from the chain rule

$$\mathbf{F}'(t) = \mathbf{G}'(\varphi(t))\varphi'(t),$$

and the fact that neither $\mathbf{F}'$ nor $\mathbf{G}'$ vanishes. Conversely, if $\mathbf{G}(u)$ parametrizes a curve and if $u = \phi(t)$ is a smooth, monotonically increasing function, then setting $\mathbf{F}(t) = \mathbf{G}(\phi(t)) = \mathbf{G}(u)$ defines a *reparametrization* of the original curve.

Geometric Invariants

We are now ready to start computing the geometric invariants of parametric curves. We will begin with the arclength, which is the integral of the speed. This is a good starting point, because we can use the arclength function to reparametrize a smooth curve by a parametrization that has constant unit speed; doing so will make the rest of the discussion simpler.

Arclength

We define the arclength to be the integral of the speed. More precisely, the *arclength* on the curve $\mathbf{r}(t)$ between $t = a$ and $t = b$ is

$$\int_a^b v(t)\,dt = \int_a^b \|\mathbf{v}(t)\|\,dt.$$

Since the speed is nonnegative, the *arclength function*

$$s(t) = \int_a^t v(u)\,du$$

is a monotonically increasing, smooth function of t. Notice that s depends on t in the manner described by the Fundamental Theorem of Calculus, so we can compute the derivative by the formula

$$\frac{ds}{dt} = v(t) = \|\mathbf{v}(t)\|.$$

Next, let's check that the arclength invariant is independent of parametrization. This follows from a simple change of variable argument:

$$\int_c^d \|\mathbf{G}'(u)\|\,du = \int_a^b \|\mathbf{G}'(\varphi(t))\|\|\varphi'(t)\|\,dt$$

$$= \int_a^b \|\mathbf{F}'(t)\|\,dt.$$

We can reparametrize our curve by using arclength itself as the parameter. This may be computationally very difficult, but theoretically it will make our ensuing arguments much, much easier. Thus, reparametrizing the curve in terms of arclength has the net effect of making the speed function equal to the constant 1.

So, given any (piecewise) smooth curve, it is no loss of generality to assume that it is parametrized by its arclength.

Let's see how this idea works in the case of the helix. Since the speed is equal to $\sqrt{2}$, its integral is the arclength function $s = \sqrt{2}\,t$. In this case, we can easily solve for t and write $t = s/\sqrt{2}$. Thus, we can reparametrize the helix as follows:

```
In[7]:= unitHelix = helix /. t -> s/Sqrt[2]
```

$$\text{Out[7]} = \left\{ \cos\left[\frac{s}{\sqrt{2}}\right], \sin\left[\frac{s}{\sqrt{2}}\right], \frac{s}{\sqrt{2}} \right\}$$

By finding the length of the velocity vector, we can check that the parametrization by arclength gives a curve whose speed always equals 1.

```
In[8]:= newSpeed = vectorLength[D[unitHelix, s]]
```

```
Out[8]= 1
```

The Frenet Frame

Until further notice, our curves will be parametrized by arclength. So, we will write

$$\mathbf{r} = \mathbf{r}(s), \qquad s \in I,$$

where

$$\left\| \frac{d\mathbf{r}}{ds} \right\| = \|\mathbf{r}'(s)\| \equiv 1, \qquad s \in I.$$

Such curves are called *unit-speed curves*. As we just saw, a smooth curve depending on any parameter t can always be converted into a unit-speed curve by reparametrizing it in terms of its arclength function $s(t)$.

The velocity vector

$$\mathbf{T}(s) = \frac{d\mathbf{r}}{ds} = \frac{d\mathbf{r}}{dt}\frac{dt}{ds} = \frac{1}{\|\mathbf{v}\|}\frac{d\mathbf{r}}{dt} = \frac{\mathbf{v}}{\|\mathbf{v}\|}$$

is a *unit tangent* vector to the curve at $\mathbf{r}(s)$. In this case, it is perpendicular to the acceleration vector. This fact is a simple consequence of a principle that we shall use repeatedly: *a vector-valued function with constant magnitude is perpendicular to its derivative.* To see this, simply note that differentiating the equation

$$\mathbf{F}(t) \cdot \mathbf{F}(t) \equiv c$$

yields

$$2\,\mathbf{F}(t) \cdot \mathbf{F}'(t) = 0.$$

Now apply the computation to $\mathbf{T}(s)$ itself. The derivative is perpendicular to the unit tangent $\mathbf{T}(s)$, so it must be normal to the curve. Adjusting its length, we get a *unit normal* vector

$$\mathbf{N}(s) = \frac{\mathbf{T}'(s)}{\|\mathbf{T}'(s)\|},$$

which satisfies

$$\mathbf{T}(s) \cdot \mathbf{N}(s) = 0.$$

The vector **N** is often called the *principal normal*.

The vectors **T** and **N** determine a plane, called the *osculating plane*. We shall explain the term below. (It can be shown that if $\mathbf{N} \neq \mathbf{0}$, then it points "into the concavity" when the curve is projected onto the osculating plane.) Of course, we want the osculating plane to exist, so we shall assume tacitly that the derivative $\mathbf{T}'$ vanishes at only finitely many points in any bounded interval. Finally, we set

$$\mathbf{B}(s) = \mathbf{T}(s) \times \mathbf{N}(s)$$

and call it the unit *binormal* vector. It is perpendicular to both **T** and **N**. The three vectors together $\mathbf{T}(s)$, $\mathbf{N}(s)$, $\mathbf{B}(s)$ constitute a *Frenet frame* (named after the mid-nineteenth century French mathematician Frenet).

How does the Frenet frame depend on the parametrization? We saw above that the unit tangent vector **T** is the same whether we use the parametrization by t or the parametrization by arclength s. (In fact, the unit tangent only depends on the orientation.) Similarly, we have

$$\frac{d\mathbf{T}}{ds} = \frac{d\mathbf{T}}{dt}\frac{dt}{ds} = \frac{1}{\|\mathbf{v}\|}\frac{d\mathbf{T}}{dt}.$$

In other words, the two vector derivatives differ by a positive scalar function. So, the corresponding unit normal vectors are identical. Clearly, the binormal also turns out to be independent of the parametrization.

Returning to the helix, we will compute its Frenet frame using the parametrization by arclength. Since we'll need to use the `CrossProduct` command, we start by loading the `Calculus`VectorAnalysis`` package. We also define another command that we will use repeatedly to compute unit vectors.

```
In[9]:=  <<Calculus`VectorAnalysis`
```

```
In[10]:= unitVector[x_] := Simplify[x/vectorLength[x]]
```

```
In[11]:= UT = D[unitHelix, s]
```

$$\text{Out[11]=} \quad \left\{ -\frac{\text{Sin}\left[\frac{s}{\sqrt{2}}\right]}{\sqrt{2}}, \frac{\text{Cos}\left[\frac{s}{\sqrt{2}}\right]}{\sqrt{2}}, \frac{1}{\sqrt{2}} \right\}$$

```
In[12]:= UN = unitVector[D[UT, s]]
```

$$\text{Out[12]=} \quad \left\{ -\text{Cos}\left[\frac{s}{\sqrt{2}}\right], -\text{Sin}\left[\frac{s}{\sqrt{2}}\right], 0 \right\}$$

In[13]:= **UB = Simplify[CrossProduct[UT, UN]]**

Out[13]= $\left\{ \dfrac{\mathrm{Sin}\left[\frac{s}{\sqrt{2}}\right]}{\sqrt{2}}, -\dfrac{\mathrm{Cos}\left[\frac{s}{\sqrt{2}}\right]}{\sqrt{2}}, \dfrac{1}{\sqrt{2}} \right\}$

Once we've computed UT, UN, and UB as functions of s, it is possible to plot them at any point on the curve by using **Graphics** and **Line** to draw the three Frenet vectors emanating from that point. A program for doing this is on the accompanying computer disk. The following figure shows two Frenet frames for the helix at two different points:

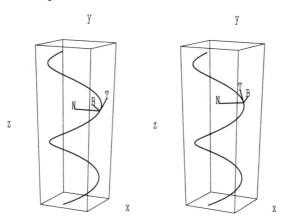

Note how the frame rotates as one moves along the curve. It is sometimes useful to think of the Frenet frame as a moving coordinate system specially adapted to the curve. You can see this better by producing a sequence of pictures of Frenet frames using the program on the disk. If you then highlight the right cell brackets of these pictures and click on **Animate Selected Graphics** in the **Cell** menu, you can see a movie illustrating this point.

Curvature and Torsion

The *curvature* measures, at any point, how much the curve is bending. Said another way, it measures the failure of the curve to be linear. It is defined by the formula

$$\kappa(s) = \|\mathbf{T}'(s)\|, \qquad s \in I.$$

Thus $\mathbf{T}'(s) = \kappa(s)\mathbf{N}(s)$ by definition. The *torsion* measures how much the curve is twisting, or the failure of the curve to be planar. The torsion is uniquely specified by the equation

$$\mathbf{B}'(s) = -\tau(s)\mathbf{N}(s).$$

Indeed, we observe that $\mathbf{B}'(s) \cdot \mathbf{B}(s) = 0$, since $\mathbf{B}(s)$ has constant length. Moreover, $\mathbf{B}'(s) \cdot \mathbf{T}(s) = 0$ (which follows by differentiating the equation $\mathbf{B}(s) \cdot \mathbf{T}(s) = 0$).

Hence the vectors $\mathbf{B}'$ and $\mathbf{N}$ must be parallel, so they differ by a scalar.

We illustrate these ideas by computing the curvature and torsion of the helix.

```
In[14]:= helcurv = vectorLength[D[UT, s]]
```

$$\text{Out[14]} = \frac{1}{2}$$

Thus, the helix has constant curvature, $\kappa = 1/2$.

```
In[15]:= D[UB, s]
```

$$\text{Out[15]} = \left\{ \frac{1}{2}\text{Cos}\left[\frac{s}{\sqrt{2}}\right], \frac{1}{2}\text{Sin}\left[\frac{s}{\sqrt{2}}\right], 0 \right\}$$

We observe that the derivative of the binormal is negative one-half of the unit normal vector, so the torsion of the helix is also constant, $\tau = 1/2$.

The three vectors in the Frenet frame form a right-handed frame field of mutually perpendicular unit vectors, traveling along the curve. The definitions of curvature and torsion depend on the fact that the derivatives of $\mathbf{T}$ and $\mathbf{B}$ are parallel to $\mathbf{N}$. Can we say anything about the derivative of $\mathbf{N}$?

Theorem 1 (Frenet Formulas). *If $\mathbf{r}$ is a smooth unit-speed curve with positive curvature κ and torsion τ, then we have the following family of first-order differential equations:*

$$\begin{aligned}
\mathbf{T}' &= & \kappa\mathbf{N}, \\
\mathbf{N}' &= -\kappa\mathbf{T} & +\tau\mathbf{B}, \\
\mathbf{B}' &= & -\tau\mathbf{N}.
\end{aligned}$$

Proof. Only the second differential equation remains to be verified. In fact $\mathbf{N}'$ can be expanded in terms of the frame field $\{\mathbf{T}, \mathbf{N}, \mathbf{B}\}$:

$$\mathbf{N}' = (\mathbf{N}' \cdot \mathbf{T})\mathbf{T} + (\mathbf{N}' \cdot \mathbf{N})\mathbf{N} + (\mathbf{N}' \cdot \mathbf{B})\mathbf{B}.$$

Differentiating $\mathbf{N} \cdot \mathbf{T} = 0$, we see that

$$\mathbf{N}' \cdot \mathbf{T} = -\mathbf{N} \cdot \mathbf{T}' = -\mathbf{N} \cdot \kappa\mathbf{N} = -\kappa.$$

Also $\mathbf{N} \cdot \mathbf{N}' = 0$ since $\mathbf{N}$ has constant length. Finally, we have

$$\mathbf{N}' \cdot \mathbf{B} = -\mathbf{N} \cdot \mathbf{B}' = -\mathbf{N} \cdot (-\tau\mathbf{N}) = \tau. \qquad \blacksquare$$

The Frenet formulas in Theorem 1 have been simplified slightly because we are using the parametrization by arclength. In practice, it may be extremely difficult to find a formula for the arclength function, so it may be computationally impossible to reparametrize a given curve into a unit-speed curve. We need to have formulas for the geometric invariants expressed in terms of variable-speed parametrizations. We present these formulas in the next paragraph without justification.

If **F** is any vector-valued function of the parameter on a curve, then we can think of **F** as a function of s or of t. By the chain rule, we have

$$\frac{d\mathbf{F}}{dt} = \frac{d\mathbf{F}}{ds}\frac{ds}{dt} = v(t)\frac{d\mathbf{F}}{ds}.$$

Applying this formula to the vectors in the Frenet frame, we can rewrite the Frenet formulas for any parametric curve in the form

$$\begin{aligned}
\mathbf{T}' &= & \kappa v\mathbf{N}, \\
\mathbf{N}' &= -\kappa v\mathbf{T} & +\tau v\mathbf{B}, \\
\mathbf{B}' &= & -\tau v\mathbf{N}.
\end{aligned}$$

The velocity and acceleration vectors are

$$\mathbf{v} = v\mathbf{T}, \qquad \mathbf{a} = \frac{dv}{dt}\mathbf{T} + \kappa v^2\mathbf{N}.$$

It is useful to keep in mind that **v** and **a** both lie in the osculating plane, and that **N** and **a** are on the same side of **T**, as in the following picture:

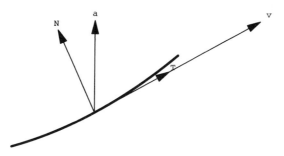

Here **B** would point straight out of the page. If we compute $\mathbf{v} \times \mathbf{a}$, the term involving dv/dt cancels, so the curvature and torsion are given by

$$\kappa = \frac{\|\mathbf{v} \times \mathbf{a}\|}{\|\mathbf{v}\|^3}, \qquad \tau = \frac{(\mathbf{v} \times \mathbf{a}) \cdot \mathbf{a}'}{\|\mathbf{v} \times \mathbf{a}\|^2}.$$

It may be easier computationally to compute the Frenet frame by using only the velocity and acceleration vectors, namely

$$\mathbf{T} = \frac{\mathbf{v}}{\|\mathbf{v}\|}, \qquad \mathbf{B} = \frac{\mathbf{v} \times \mathbf{a}}{\|\mathbf{v} \times \mathbf{a}\|}, \qquad \mathbf{N} = \mathbf{B} \times \mathbf{T}.$$

Differential Geometry of Curves

The Frenet formulas have many uses in the subject called *differential geometry*. Differential geometry involves a sophisticated study of the uses of calculus to penetrate the mysteries of geometric objects. We shall content ourselves here with using the Frenet formulas to solve some elementary problems:

(1) We know that the best linear approximation to a curve at a point is the tangent vector. What is the best approximation by a circle?

(2) Give a criterion to determine when a curve is planar.

(3) Give a condition to determine when a curve is spherical.

(4) Give a condition to determine when a curve is helical. When is it a circular helix?

(5) Show that all the geometric invariants we have defined are independent of parametrization.

We shall give solutions to problems (1)–(4) in terms of the geometric invariants we have developed. To simplify the discussion, we will continue to assume that the curve is parametrized by arclength. Before beginning, we call your attention to the computation of the geometric invariants for the helix. In particular, we note that the speed, curvature, and torsion are all nonzero, but constant. Now we begin to solve the problems.

The Osculating Circle

The circle that best approximates the curve near a point must be one that is tangent to the curve at the point in question. But there are many such circles. We pick out the circle that has the same curvature as the curve and has center along the ray pointing in the direction of the principal normal vector. Since a circle of radius ρ has constant curvature $\kappa = 1/\rho$, that can be done by selecting the circle of radius $1/\kappa(s)$, centered at the point $\mathbf{r}(s) + (1/\kappa(s))\mathbf{N}(s)$. The radius of this circle, $\rho = 1/\kappa$, is called the *radius of curvature* of the curve at the appropriate point. The circle "kisses" the curve accurately to second order, thus is given the name *osculating circle* (from the Latin word for "kissing"). The plane it lies in (i.e., the one determined by $\mathbf{T}(s)$ and $\mathbf{N}(s)$) is therefore referred to as the osculating plane. The curve determined by the centers of all osculating circles is called the *evolute* of the curve. The illustration on the next page shows one of the osculating circles to the helix. You can see how it was made by looking at the disk accompanying this book. Using the program given there, you can also do an animation to show how the osculating circle moves as you go from point to point on the curve.

Plane Curves

If we think of plane curves as those which bend but do not twist, then the following theorem is perfectly natural.

Theorem 2. *A unit-speed curve is planar if and only if its torsion is zero.*

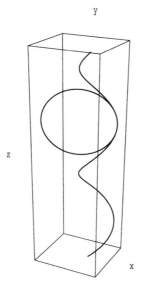

Proof. If the curve is planar, then the binormal vector is constant and thus has zero derivative. Therefore the torsion, which is, up to scalar, the length of the derivative of the binormal, must vanish. Conversely, suppose $\tau(s) = 0$ for all s. Then $\mathbf{B}'(s) = 0$. Hence $\mathbf{B}(s)$ is a constant vector, say $\mathbf{B}$. Then in fact the curve lies in the plane through $\mathbf{r}(0)$ perpendicular to $\mathbf{B}$. To see that, set $g(s) = (\mathbf{r}(s) - \mathbf{r}(0)) \cdot \mathbf{B}$. Then $g'(s) = \mathbf{T}(s) \cdot \mathbf{B} = 0$. But $g(0) = 0$. Therefore g is identically zero, which proves the assertion. ∎

Spherical Curves

Suppose that $\mathbf{r}$ lies on a sphere of radius r centered at $\mathbf{c}$. Then the vector function $\mathbf{r} - \mathbf{c}$ points from the center of the sphere to the curve and therefore must be perpendicular to the unit tangent vector. Hence we can write $\mathbf{r} = \mathbf{c} + u(s)\mathbf{N}(s) + w(s)\mathbf{B}(s)$, for some as yet undetermined functions u and w. Then we differentiate and use the Frenet formulas

$$\mathbf{T}(s) = \mathbf{r}'(s)$$
$$= u'(s)\mathbf{N}(s) + u(s)(-\kappa(s)\mathbf{T}(s) + \tau(s)\mathbf{B}(s))$$
$$+ w'(s)\mathbf{B}(s) + w(s)(-\tau(s)\mathbf{N}(s)).$$

Equating coefficients in the Frenet frame, we find that

$$u(s) = -\rho(s) \quad \text{and} \quad w(s) = -\rho'(s)\sigma(s),$$

where by definition $\rho = 1/\kappa$ and $\sigma = 1/\tau$. Hence $\rho^2 + (\rho'\sigma)^2 = r^2$ is constant.

In fact the converse is also true. That is, if the quantity $\rho^2 + (\rho'\sigma)^2$ is constant, say

$$\rho^2 + (\rho'\sigma)^2 = r^2, \qquad (*)$$

and if $\rho' \neq 0$, then $\mathbf{r}$ lies on a sphere of radius r. To see that, consider the curve $\mathbf{c}(s) = \mathbf{r}(s) + \rho(s)\mathbf{N}(s) + \rho'(s)\sigma(s)\mathbf{B}(s)$. It suffices to show that the curve is constant. To do this, simply compute $\mathbf{c}'$ and employ the Frenet formulas and equation $(*)$; we leave the computation to the reader. (At one point in the computation, we must divide by ρ', which explains why we assume $\rho' \neq 0$. Otherwise, a helix would supply a counterexample.)

Helical Curves

Consider the right circular helix from our first example. It is clear that at every point on the helix, the unit tangent vector makes a constant angle with respect to the z-axis. With that in mind, we make the definition: a curve $\mathbf{r}$ is called a *cylindrical helix* if there is a fixed unit vector $\mathbf{u}$ such that $\mathbf{T}(s) \cdot \mathbf{u}$ is constant. Then we have

Theorem 3. *A curve with nonvanishing curvature is a cylindrical helix if and only if the ratio τ/κ is constant.*

Proof. If $\mathbf{r}$ is a cylindrical helix, then $\mathbf{T} \cdot \mathbf{u} = \cos\theta$, where $\mathbf{u}$ is a fixed unit vector and θ is a fixed angle. But then

$$0 = (\mathbf{T} \cdot \mathbf{u})' = \mathbf{T}' \cdot \mathbf{u} = \kappa \mathbf{N} \cdot \mathbf{u}.$$

Since κ is positive (by hypothesis), we find that $\mathbf{N} \cdot \mathbf{u} = 0$. Hence $\mathbf{u}$ lies in the plane determined by the tangent and binormal vectors. Thus

$$\mathbf{u} = \cos\theta\, \mathbf{T}(s) + \sin\theta\, \mathbf{B}(s).$$

We differentiate and make use of the Frenet formulas to obtain

$$(\kappa\cos\theta - \tau\sin\theta)\mathbf{N} = 0.$$

Therefore $\tau/\kappa = \cot\theta$, a constant.

Conversely, suppose τ/κ is constant, say $\cot\theta$ for some choice of θ. Set $\mathbf{U}(s) = \cos\theta\,\mathbf{T}(s) + \sin\theta\,\mathbf{B}(s)$. Differentiating, we find $\mathbf{U}'(s) = (\kappa(s)\cos\theta - \tau(s)\sin\theta)\mathbf{N}(s)$. By assumption, the latter is zero. Hence $\mathbf{U}$ is constant. But then $\mathbf{U}(s) = \mathbf{U}(0) = \cos\theta\mathbf{T}(0) + \sin\theta\mathbf{B}(0)$. Hence $\mathbf{T}(s) \cdot \mathbf{U} = \cos\theta$, for any s. ∎

Now in the case of a right circular helix, not only is the quotient τ/κ constant, each of the quantities κ and τ is constant. To see that that property characterizes right circular helices (and solves problem (4) completely), we must proceed to the last problem.

Congruence

Here we get a little too heavily into differential geometry to give a complete proof; we'll content ourselves with an informal discussion. A vector-valued function of three variables $\mathbf{F}(x, y, z) = (f(x, y, z), g(x, y, z), h(x, y, z))$ can be thought of as a transformation of space $\mathbf{F} : \mathbf{R}^3 \to \mathbf{R}^3$. Such a transformation is called an *isometry* if it preserves distance, that is, $d(\mathbf{F}(\mathbf{p}), \mathbf{F}(\mathbf{q})) = d(\mathbf{p}, \mathbf{q})$, for every pair of points $\mathbf{p}$ and $\mathbf{q}$ in three-space, where d is the Euclidean distance function. Examples of isometries include:

- parallel translations;
- rotations about an axis; and
- reflections in a plane.

It is a fact that any isometry is just a combination of these three types of transformations. The first two are examples of *orientation-preserving* transformations; the latter is *orientation-reversing*. We won't give precise definitions of these terms; suffice it to say that if we fix a right-handed frame (i.e., a collection $\{e_1, e_2, e_3\}$ consisting of three mutually perpendicular unit vectors satisfying $e_1 \cdot (e_2 \times e_3) = 1$), then an orientation-preserving isometry will convert it into another right-handed frame, while an orientation-reversing one will make it left-handed (i.e., $e_1 \cdot (e_2 \times e_3) = -1$). We say that two curves $\mathbf{r}_1$, $\mathbf{r}_2$ are *congruent* if there is an isometry of three-space that transforms one curve into the other.

Here is the relationship between the geometric invariants of two congruent curves. Let $\mathbf{r}_1$ be a smooth curve, $\mathbf{F}$ an isometry, and $\mathbf{r}_2 = \mathbf{F} \circ \mathbf{r}_1$. If $\mathbf{F}$ is a translation, then the Frenet frame, speed, curvature, and torsion are exactly the same. (Translations correspond to adding a constant vector to the parametrization; the constant disappears the instant that you take a derivative to compute the velocity.) If $\mathbf{F}$ consists of rotations or reflections, then

$$
\begin{aligned}
\mathbf{T}_2 &= \mathbf{F}(\mathbf{T}_1), & \kappa_2 &= \kappa_1, \\
\mathbf{N}_2 &= \mathbf{F}(\mathbf{N}_1), & \tau_2 &= \mathrm{sgn}(\mathbf{F}) \tau_1, \\
\mathbf{B}_2 &= \mathrm{sgn}(\mathbf{F}) \, \mathbf{F}(\mathbf{B}_1),
\end{aligned}
$$

where $\mathrm{sgn}(\mathbf{F})$ is ± 1 according as $\mathbf{F}$ is orientation-preserving or reversing. Now comes the basic result

Theorem 4. *If $\mathbf{r}_1$ and $\mathbf{r}_2$ have the property that $\kappa_1 = \kappa_2$ and $\tau_1 = \pm\tau_2$, then the two curves are congruent.*

Proof. First assume $\tau_1 = \tau_2$. Consider the two frames $\{\mathbf{T}_1(0), \mathbf{N}_1(0), \mathbf{B}_1(0)\}$ and $\{\mathbf{T}_2(0), \mathbf{N}_2(0), \mathbf{B}_2(0)\}$. There must be an orientation-preserving isometry $\mathbf{F}$ that carries the first frame into the second. Set $\mathbf{r}_* = \mathbf{F} \circ \mathbf{r}_1$. Then by the preceding formulas, the two curves $\mathbf{r}_2$ and $\mathbf{r}_*$ have the same geometric data at the origin. Now consider the scalar-valued function

$$h = \mathbf{T}_* \cdot \mathbf{T}_2 + \mathbf{N}_* \cdot \mathbf{N}_2 + \mathbf{B}_* \cdot \mathbf{B}_2.$$

Since $\mathbf{T}_*$ and $\mathbf{T}_2$ are both unit vectors, we have $\mathbf{T}_* \cdot \mathbf{T}_2 \le 1$. Furthermore, the dot product equals 1 precisely when $\mathbf{T}_* = \mathbf{T}_2$. Similarly with the other two frame components. Thus it suffices to show that $h(s) \equiv 3$. By the definition of $\mathbf{r}_*$, we have $h(0) = 3$. Now it suffices to show $h' = 0$. But as usual, this follows by differentiating, and then using the Frenet formulas together with the equality of the respective curvatures and torsions.

We have proven, in particular, that $\mathbf{T}_* = \mathbf{T}_2$. Thus, the components of the position vectors have the same derivative. By elementary calculus, they can differ by at most a constant. This says that $\mathbf{r}_* = \mathbf{r}_2 + \mathbf{c}$. But the curves agree at $s = 0$, hence $\mathbf{c} = \mathbf{0}$. Therefore, the fixed isometry $\mathbf{F}$ transforms the curve $\mathbf{r}_1$ into the curve $\mathbf{r}_2$, that is the two curves are congruent.

The case $\tau_1 = -\tau_2$ is handled similarly, except that an orientation-reversing transformation is employed. ∎

The dispensation of problem (4) comes swiftly now. Right circular helices have constant curvature and torsion. Since these quantities determine the nature of a curve (up to congruence), the constancy of them characterize helices.

It is also a simple matter to augment Theorem 4 to the variable-speed case.

Corollary 5.   *Two curves which have the same speed, curvature, and torsion (the latter up to sign) must be congruent.*

The proof is very easy. The unit-speed reparametrizations of the two curves have matching curvatures and torsions, therefore are congruent by Theorem 4. The original curves are then easily seen to be congruent. We leave the details to the reader.

Two More Examples

Our last task is to work out the geometric invariants for the remaining examples. For this purpose, we will use the following functions, whose argument should be thought of as the position vector, written as a list of expressions in the parameter t. They also assume we have already loaded the `Calculus`VectorAnalysis`` package, and defined the commands `vectorLength` and `unitVector` as above.

```
In[16]:= Clear[UT, UN, UB]
```

```
In[17]:= velocity[r_] := Simplify[D[r, t]]
         acceleration[r_] := Simplify[D[r, {t, 2}]]
         speed[r_] := vectorLength[velocity[r]]
         UT[r_] := unitVector[velocity[r]]
         UN[r_] := unitVector[D[UT[r], t]]
         UB[r_] := Simplify[CrossProduct[UT[r], UN[r]]]
```

The curvature and torsion functions are slightly more complicated to define. The programs here use the definitions coming directly from the Frenet formulas for variable-speed curves presented on page 43; more efficient programs would use the alternate formulas in terms of the velocity and acceleration on the same page.

```
In[23]:= curvature[r_] := Simplify[
           vectorLength[D[UT[r], t]]/speed[r]]
         torsion[r_] := tau /. Simplify[First[
           Solve[D[UB[r], t] == -tau*speed[r]*UN[r], tau]]]
```

Now we are ready to look at the examples.

The Astroid

The astroid was defined parametrically by the following equations:

```
In[25]:= astroid = {Cos[t]^3, Sin[t]^3, Cos[2t]}
```

$$Out[25]= \{Cos[t]^3, Sin[t]^3, Cos[2t]\}$$

Here is a graph of the astroid. We immediately notice the most significant geometric features of the curve: there are four cusps where the curve fails to be smooth. When we compute the geometric invariants, we should look for evidence of the cusps.

```
In[26]:= ParametricPlot3D[Evaluate[astroid], {t, 0, 2Pi},
           ViewPoint->{1.5, 3, 1}];
```

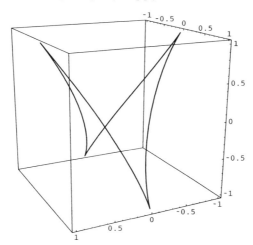

We compute the velocity, speed, and acceleration of the astroid.

In[27]:= **velocity[astroid]**

Out[27]= $\{-3\text{Cos}[t]^2\text{Sin}[t], 3\text{Cos}[t]\text{Sin}[t]^2, -2\text{Sin}[2t]\}$

In[28]:= **speed[astroid]**

Out[28]= $\dfrac{5}{2}\sqrt{\text{Sin}[2t]^2}$

In[29]:= **acceleration[astroid]**

Out[29]= $\left\{-\dfrac{3}{4}(\text{Cos}[t] + 3\text{Cos}[3t]), -\dfrac{3}{4}(\text{Sin}[t] - 3\text{Sin}[3t]), -4\text{Cos}[2t]\right\}$

Using Out[28], we see that the speed of the astroid is zero precisely when $\sin(2t) = 0$, which occurs when $t = 0$, $\pi/2$, π, or $3\pi/2$. Thus, we have identified the locations of the four cusps that we saw in the graph.

Our next task is to compute the Frenet frame for the astroid.

In[30]:= **UT[astroid]**

Out[30]= $\left\{-\dfrac{6\text{Cos}[t]^2\text{Sin}[t]}{5\sqrt{\text{Sin}[2t]^2}}, \dfrac{6\text{Cos}[t]\text{Sin}[t]^2}{5\sqrt{\text{Sin}[2t]^2}}, -\dfrac{4}{5}\text{Csc}[2t]\sqrt{\text{Sin}[2t]^2}\right\}$

In[31]:= **UN[astroid]**

Out[31]= $\left\{\dfrac{2\text{Cos}[t]\text{Sin}[t]^2}{\sqrt{\text{Sin}[2t]^2}}, \dfrac{2\text{Cos}[t]^2\text{Sin}[t]}{\sqrt{\text{Sin}[2t]^2}}, 0\right\}$

In[32]:= **UB[astroid]**

Out[32]= $\left\{\dfrac{4\text{Cos}[t]}{5}, \dfrac{4\text{Sin}[t]}{5}, -\dfrac{3}{5}\right\}$

Finally, we compute the curvature and the torsion of the astroid.

In[33]:= **curvature[astroid]**

Out[33]= $\dfrac{6}{25\sqrt{\text{Sin}[2t]^2}}$

In[34]:= **torsion[astroid]**

Out[34]= $\dfrac{4}{25}\text{Csc}[t]\text{Sec}[t]$

The curvature and torsion are also undefined at the cusps, where $t = 0$, $\pi/2$, π, or $3\pi/2$.

The Cycloid

The cycloid is a plane curve, defined parametrically by the following equations:

```
In[35]:= cyc = {t - Sin[t], 1 - Cos[t]}
```

Out[35]= $\{t - \text{Sin}[t], 1 - \text{Cos}[t]\}$

```
In[36]:= ParametricPlot[Evaluate[cyc], {t, 0, 4Pi},
         AspectRatio -> Automatic];
```

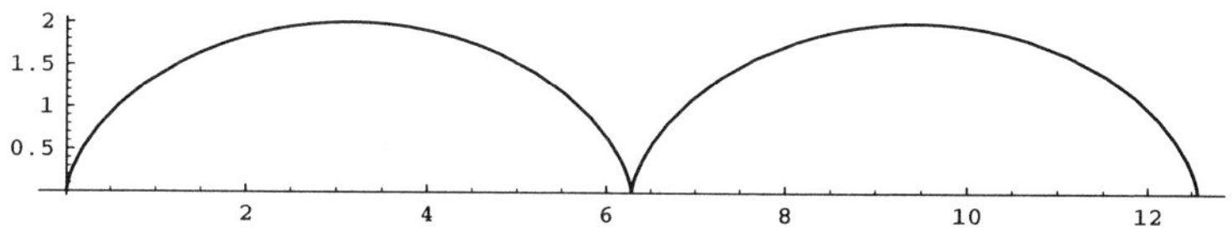

In the graph of the cycloid, we have used `AspectRatio -> Automatic` to cause *Mathematica* to display the graph using the same scale on both axes. Since the cycloid is defined as the trace of a point on the edge of a rolling wheel, it is only reasonable to insist that the graph should contain a realistic picture. (Try the graph using the default setting for `AspectRatio` to see what happens.)

Before proceeding, we need to discuss a difference between plane curves and space curves. The formulas (and the *Mathematica* programs) for velocity, speed, acceleration, unit tangent vector, principal normal vector, and curvature work equally well for both kinds of curves. The formulas (and programs) for the binormal vector and for torsion only work for space curves, since they involve a cross product. Since we know the values of these items for plane curves, we could simply avoid the computation. Another approach is to add a zero component at the end of the plane parametrization.

```
In[37]:= cycloid = Join[cyc, {0}]
```

Out[37]= $\{t - \text{Sin}[t], 1 - \text{Cos}[t], 0\}$

```
In[38]:= velocity[cycloid]
```

Out[38]= $\{1 - \text{Cos}[t], \text{Sin}[t], 0\}$

```
In[39]:= acceleration[cycloid]
```

Out[39]= $\{\text{Sin}[t], \text{Cos}[t], 0\}$

```
In[40]:= speed[cycloid]
```

Out[40]= $\sqrt{2 - 2\text{Cos}[t]}$

The speed of the cycloid is zero when $\cos(t) = 1$, which happens when $t = 0, 2\pi$, 4π, etc. These are the cusps where the curve is not smooth. We can now proceed to compute the Frenet frame for the cycloid.

In[41]:= **UT[cycloid]**

Out[41]= $\left\{ \sqrt{\operatorname{Sin}\left[\frac{t}{2}\right]^2}, \dfrac{\operatorname{Sin}[t]}{\sqrt{2 - 2\operatorname{Cos}[t]}}, 0 \right\}$

In[42]:= **UN[cycloid]**

Out[42]= $\left\{ \operatorname{Cot}\left[\frac{t}{2}\right] \sqrt{\operatorname{Sin}\left[\frac{t}{2}\right]^2}, -\sqrt{\operatorname{Sin}\left[\frac{t}{2}\right]^2}, 0 \right\}$

In[43]:= **UB[cycloid]**

Out[43]= $\{0, 0, -1\}$

Since the cycloid is a plane curve, we knew that the binormal vector would point in the z-direction. How could we have predicted the negative sign? What does it mean?

In[44]:= **curvature[cycloid]**

Out[44]= $\dfrac{1}{2\sqrt{2 - 2\operatorname{Cos}[t]}}$

The curvature of the cycloid blows up at the cusps.

In[45]:= **torsion[cycloid]**

Out[45]= 0

Unsurprisingly, the torsion of a plane curve is zero.

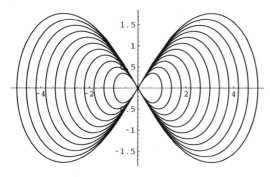

Problem Set C

CURVES

When working on this problem set, you may use the programs from Chapter 3, *Geometry of Curves*. It is not necessary to retype them; they may be found on the disk accompanying this book.

1. Each part of this problem contains parametric equations for a curve. You should:

- Graph each curve.
- Compute the Frenet frame, speed, curvature, and torsion.
- Describe any points where the curvature or torsion is undefined, and relate those points to geometric properties of the curve as revealed by the graph.

Keep the following points in mind as you work the problems:

- Variable-speed curves may require you to use different formulas than unit-speed curves.
- Plane curves may require different treatment than space curves.
- If you expect to do more than one of these problems, then you should consider using programs like those presented in Chapter 3.

   (a) The cardioid:

$$\mathbf{r}(t) = 2\cos t(1 + \cos t)\,\mathbf{i} + 2\sin t(1 + \cos t)\,\mathbf{j}, \quad 0 \le t \le 2\pi.$$

(b) The twisted cubic curve:

$$\mathbf{r}(t) = (t, t^2, t^3), \quad -2 \le t \le 2.$$

(c) The tractrix:

$$\mathbf{r}(t) = \begin{cases} e^{-t}\mathbf{i} + \int_0^t \sqrt{1 - e^{-2u}}\, du\, \mathbf{j}, & t \ge 0, \\ e^t\mathbf{i} + \int_0^t \sqrt{1 - e^{2u}}\, du\, \mathbf{j}, & t \le 0. \end{cases}$$

You may restrict your computations of the geometric data to one of the branches of the curve, but be sure to include both branches when you plot it.

(d) A hyperbolic helix:

$$\mathbf{r}(t) = \left(\sqrt{1 + \frac{t^2}{2}}, \frac{t}{\sqrt{2}}, \operatorname{arcsinh} \frac{t}{\sqrt{2}} \right), \quad t \in \mathbf{R}.$$

In this case, make the change of variable $t = \sqrt{2} \sinh u$ in the coordinates of $\mathbf{r}(t)$ and see if you can justify the name of the curve.

(e) A portion of a pseudo-circular helix:

$$\mathbf{r}(t) = \left(\frac{1}{3}(1 + t)^{3/2}, \frac{1}{3}(1 - t)^{3/2}, \frac{t}{\sqrt{2}} \right), \quad -1 \le t \le 1.$$

In this case, show that when the curve is projected onto the x-y plane, its image is the portion of the curve $x^{2/3} + y^{2/3} = 2/3^{2/3}$ that lies in the first quadrant. We say that a curve $x^a + y^a = c$, for a and c positive, is a *pseudo-circle*.

(f) Viviani's curve:

$$\mathbf{r}(t) = (1 + \cos t, \sin t, 2\sin(t/2)), \quad -\pi \le t \le \pi.$$

This curve is the intersection of the sphere $x^2 + y^2 + z^2 = 4$ with the cylinder $(x - 1)^2 + y^2 = 1$. You may use the program from Chapter 2 to graph it.

(g) The catenary:

$$\mathbf{r}(t) = (t, \cosh t).$$

(h) The limaçon:

$$\mathbf{r}(t) = (2\cos t + 1)(\cos t, \sin t).$$

2. Find the arclength of the following curves over the indicated interval. Be prepared to use **NIntegrate** if **Integrate** fails.

   (a) $(\cos^3 t, \sin^3 t)$, $t \in [0, 2\pi]$.

   (b) $((1/3)(1 + t)^{3/2}, (1/3)(1 - t)^{3/2}, t/\sqrt{2})$, $t \in [-1, 1]$.

   (c) $\cosh t\, \mathbf{i} + \sinh t\, \mathbf{j} + t\, \mathbf{k}$, $t \in [0, 1]$.

(d) $(\sin t, \cos 2t, t^3)$, $t \in [0, \pi]$.

(e) (t, t^2, t^3), $t \in [0, 1]$.

(f) $(\arctan t, \cosh t, \ln(1 + t))$, $t \in [0, 1]$.

3. Each part of this problem contains a description of a curve as the intersection of two surfaces. Find parametric equations for each curve. Then graph the curve, and find its Frenet frame, speed, curvature, and torsion. Using the criteria presented in Chapter 3, check each curve to see if it is planar, spherical, or helical.

   (a) Intersect the saddle surface $z = xy$ with the cylinder $x^2 + y^2 = 1$.

   (b) Intersect the paraboloid $z = x^2 + y^2$ with the plane $2x + 2y + z = 2$.

   (c) Intersect the hyperboloid $x^2 + y^2 - z^2 = 1$ with the sphere $x^2 + y^2 + z^2 = 5$.

   (d) Intersect the hyperboloid $x^2 + y^2 - z^2 = 1$ with the plane $x + y = 1$.

4. Parametrize the sine curve $y = \sin x$ using the parametrization

$$\mathbf{r}(t) = (t, \sin t), \qquad t \in \mathbf{R}.$$

   (a) Using the definition in Chapter 3, show that this curve is smooth.

   (b) Compute the curvature, and find all points where the curvature is zero. What geometric property do all those points share?

   (c) Without doing any computations, explain why the torsion of this curve must be identically zero.

   (d) Rotating the plane $30°$ is an isometry, which transforms the original sine curve into the curve parametrized by

$$\mathbf{r}(t) = \left(\frac{\sqrt{3}\,t - \sin t}{2}, \frac{t + \sqrt{3}\,\sin t}{2} \right), \qquad t \in \mathbf{R}.$$

Compute the speed, curvature, and torsion of this curve, and compare them to those of the original curve.

   (e) By graphing the curve, decide if it is the graph of some function $y = f(x)$.

   (f) In single variable calculus, you studied the qualitative properties of the graphs of functions $y = f(x)$. In particular, you characterized the maxima, minima, and inflection points in terms of the vanishing of certain derivatives of the function $f(x)$. Using the earlier parts of this problem to supply examples, write a paragraph explaining why the notion of an inflection point is an intrinsic geometric property of the curve, but the notion of a maximum or minimum point is not.

5. Sometimes the formulas for curvature and torsion are so complicated that it is difficult to draw any meaningful conclusion from them. Nevertheless, you can use *Mathematica* to graph the functions κ and τ, even if their formulas are "messy". Illustrate this fact by computing κ and τ, and then graphing them together on a single two-dimensional plot. (Use the `PlotStyle` option to distinguish them.) In each case, you should also graph the curve and then reconcile your plots by relating *in detail* the features of the two-dimensional plots of the invariants to the shape of the original curves. (Note: In parts (b) and (c), you must enter the parametrization using fractions and not decimals.)

   (a) The figure eight curve, also called the lemniscate of Gerono:

   $$\mathbf{r}(t) = (\sin t, \, \cos t \sin t), \qquad 0 \le t \le 2\pi.$$

   (b) A logarithmic spiral:

   $$\mathbf{r}(t) = (e^{t/10} \cos t, \, e^{t/10} \sin t), \qquad -10\pi \le t \le 10\pi.$$

   (c) The helical logarithmic spiral:

   $$\mathbf{r}(t) = (e^{t/10} \cos t, \, e^{t/10} \sin t, t/10), \qquad -5\pi \le t \le 5\pi.$$

   (d) The portion of the intersection of the two cylinders $x^2 + y^2 = 1$ and $y^2 + z^2 = 4$ that lies above the x-y plane.

6. Compute the speed, curvature, and torsion of the curve

   $$\mathbf{r}(t) = (t + \sqrt{3} \sin t)\,\mathbf{i} + 2\cos t\,\mathbf{j} + (\sqrt{3}t - \sin t)\,\mathbf{k}.$$

   Deduce from those computations that the curve must be a helix. Confirm it by plotting the curve. Then find another helix

   $$\mathbf{R}(t) = a\cos t\,\mathbf{i} + a\sin t\,\mathbf{j} + bt\,\mathbf{k}$$

   and an isometry $\mathbf{F}$ such that $\mathbf{F}(\mathbf{r}) = \mathbf{R}$. (Hint: It's a rotation. To find it, compute the two 3×3 matrices consisting of the values of the frame fields for $\mathbf{r}$ and $\mathbf{R}$ at zero. Then find the matrix that moves one into the other. You will have to use the values of the speed, curvature, and torsion that you compute to find the correct choices of a and b. That is, find these data for $\mathbf{R}$ in terms of a and b and use `Solve`.)

7. As a point (or particle) sweeps over a curve $\mathbf{r}(t)$, the centers of its corresponding osculating circles sweep out another curve, called the *evolute* of $\mathbf{r}$. Recall that the osculating circle is the unique circle in the osculating plane (determined by $\mathbf{T}$ and $\mathbf{N}$) that is tangent to the curve and has the same curvature. Using the fact that the curvature of a circle of radius r is $1/r$, find a formula for the center

of the osculating circle in terms of $\mathbf{r}(t)$, $\mathbf{T}(t)$, $\mathbf{N}(t)$, and $\kappa(t)$. (Hint: It may be convenient to compute $\mathbf{N}$ from the formula relating it to $\mathbf{a}$ and $\mathbf{T}$.) Use that formula to obtain a parametric equation for the evolute of the following curves. In each case, plot the original curve and its evolute on the same graph. (Note: In part (c), you must enter the parametrization using fractions and not decimals.)

(a) The ellipse: $2x^2 + y^2 = 1$.

(b) The cycloid: $\mathbf{r}(t) = (t - \sin t, \ 1 - \cos t)$, $0 \le t \le 8\pi$.

(c) The helical log spiral: $\mathbf{r}(t) = (e^{t/10} \cos t, \ e^{t/10} \sin t, \ t/10)$, $0 \le t \le 6\pi$.

(d) The tractrix: Use the following parametrization:

$$\mathbf{r}(t) = (\sin t, \ \cos t + \log(\tan(t/2))), \qquad 0 \le t \le \pi.$$

Show by the transformation $u = \log(\tan(t/2))$ that the evolute of the tractrix is a catenary. (Hint: When you first compute the parametrization of the evolute, you may end up with a mixture of trig functions of both t and $t/2$. It helps to use the substitution rules {`Sin[t]` -> `2*Sin[t/2]*Cos[t/2]`}, etc., to reduce the parametrization to simpler form before substituting {`t/2` -> `ArcTan[Exp[u]]`}.)

8. In this problem, we will attempt to explain what it means to *deform* one curve into another. Intuitively, we mean something like the following. Suppose we have two smooth plane curves $\mathbf{r}_1(t)$ and $\mathbf{r}_2(t)$, each having the same initial and terminal points $\mathbf{r}_1(0) = \mathbf{r}_2(0)$, $\mathbf{r}_1(1) = \mathbf{r}_2(1)$. (Note that we conveniently assumed both curves were parametrized on the interval $0 \le t \le 1$. Because of our ability to reparametrize curves, that is no loss of generality.) We say that the curves are deformations of each other if we can *continuously* move the first into the second without cutting or tearing the curve. Here is a more precise mathematical definition. There should be a vector-valued function of two variables $\Theta(t, u)$, $0 \le t \le 1$, $0 \le u \le 1$, so that

$$\Theta(t, 0) = \mathbf{r}_1(t), \qquad \Theta(t, 1) = \mathbf{r}_2(t), \qquad t \in [0, 1].$$

It is thus clear that as u progresses from 0 to 1 the curves $t \mapsto \Theta(t, u)$ constitute the continuous deformation. For each of the following pairs of curves, find a continuous deformation between them; i.e., produce Θ, and then draw a family of curves that illustrates the deformation. Label the two curves on your picture. (Look up **Graphics** and **Text** in the Glossary or the Help Browser to see how to do the labeling in *Mathematica*.)

(a) The portion of the unit circle in the first quadrant, and the portion of the curve $x^{1/2} + y^{1/2} = 1$ in the first quadrant.

(b) The two lemniscates of Bernoulli given by

$$\mathbf{r}_j(t) = \left(\frac{j \cos t}{1 + \sin^2 t} \right) \mathbf{i} + \left(\frac{j \cos t \sin t}{1 + \sin^2 t} \right) \mathbf{j},$$

$j = 1$ or 5.

9. We have concentrated most of our attention on curves defined by parametric equations. In single-variable calculus, you studied curves in the form $y = f(x)$ and also in the form $F(x, y) = 0$. The former can be parametrized by x itself (to wit, $\mathbf{r}(x) = (x, f(x))$, but the issue of whether the locus of points (x, y) satisfying an implicit equation like $F(x, y) = 0$ can be given in parametric coordinates is quite subtle. It is bound up with the *Implicit Function Theorem*, about which we will have a little to say in Chapter 6. For the moment, let us simply investigate implicitly defined curves via the `ImplicitPlot` command. Load it via `<<Graphics'ImplicitPlot'`, then look up the syntax (by using the question mark) and use it to plot the following implicitly defined curves. Make sure to try several different ranges to be sure that you display all the interesting features of the curve.

(a) $x^2 + y^2 = 1$.

(b) $x^2 = y^3$.

(c) $x^3 - x = y^2$.

(d) $x^3 + y^3 - 3xy = \epsilon$, with $\epsilon = 0.1, 0$, and -0.1.

(e) From the pictures you obtain, explain why even if F is a "very nice" function, its locus of zeros may not be a curve in the sense in which we have been studying it. In particular, all of our parametrized curves are connected—they only have one piece. Use the Intermediate Value Theorem from one-variable calculus to show that a parametrized curve cannot have two pieces which are a fixed horizontal distance apart (e.g., all the points on one piece satisfy $x \leq 0$ and on the other piece $x \geq 1$).

10. Geometers recognize two particularly simple kinds of singularities on curves: cusps and nodes. In this problem, we will compute the geometric invariants of the simplest example of each kind of singularity.

(a) The simplest cusp, also called a semicubical parabola, is defined implicitly by the equation $y^2 = x^3$. Show that the formula

$$\mathbf{r}(t) = (t^2, t^3), \qquad t \in \mathbf{R},$$

gives a parametrization of the semicubical parabola.

(b) Graph the semicubical parabola. (You may use the implicit equations or the parametric equations.) Describe any significant geometric features of the graph.

(c) Compute the Frenet frame, speed, curvature, and torsion of the semicubical parabola. Which of these invariants, if any, can be used to detect the singularity?

(d) The simplest nodal curve is defined implicitly by the equation $y^2 = x^3 + x^2$. Show that the formula

$$\mathbf{r}(t) = (t^2 - 1, t^3 - t), \qquad t \in \mathbf{R},$$

gives a parametrization of the nodal curve.

(e) Graph the nodal curve. (You may use the implicit equations or the parametric equations.) Describe any significant geometric features of the graph.

(f) Compute the Frenet frame, speed, curvature, and torsion of the nodal curve. Which of these invariants, if any, can be used to detect the singularity?

(g) Write a paragraph explaining the differences between your results for the semicubical parabola and your results for the nodal curve.

11. It is natural to ask: given a smooth nonnegative function $\kappa(s)$, is there a smooth curve $\mathbf{r}(s)$ having $\kappa(s)$ as its curvature? As the notation suggests, there is no harm in asking that $\mathbf{r}$ be unit speed. The solution is straightforward. We only need to set

$$\theta(s) = \int_0^s \kappa(u)\, du$$

and

$$\mathbf{r}(s) = \left(\int_0^s \cos\theta(u)\, du \right) \mathbf{i} + \left(\int_0^s \sin\theta(u)\, du \right) \mathbf{j}.$$

(a) Show that the curve $\mathbf{r}(s)$ so defined indeed has curvature $\kappa(s)$.

(b) Use the formulas to find, and then graph, a curve whose curvature is

$$\kappa(s) = \frac{1}{\sqrt{1 - s^2}}, \qquad -1 < s < 1.$$

(c) Use the formulas to find, and then graph, a curve whose curvature is

$$\kappa(s) = \frac{1}{1 + s^2}, \qquad s \in \mathbf{R}.$$

(d) The method works because you (or *Mathematica*) can do the antidifferentiation. But the theory is correct even when the antidifferentiation is not possible. In that case, we can invoke the *Mathematica* command `NDSolve` to find a numerical solution to the system of *differential equations*

$$x'(s) = \cos\theta(s), \qquad y'(s) = \sin\theta(s), \qquad theta'(s) = \kappa(s).$$

Look up the syntax of `NDSolve` and use it to solve the system in the case $\kappa(s) = s + \sin s$. (Choose appropriate initial data.) Plot the solution. Note: You may find it helpful to consult the book *Differential Equations with Mathematica*, 2nd Edition, by K. Coombes, et al. (John Wiley & Sons, New York, 1997) for help in designing your `NDSolve` program.)

12. Write an essay of about 500 words entitled *How Calculus Helps Us to Understand the Geometric Nature of Curves*. In your essay, you should explain how a curve is an intrinsically geometric notion which we describe and study analytically by formulas and calculus operations. Your essay should contain only text—no mathematics formulas. It should contain two parts: in the first, explain as clearly as possible your intuitive geometric understanding of curves and their invariants; in the second, describe how those geometric notions are made precise by means of calculus. Your essay should answer, at a minimum, the following questions:

(i) What does it mean to express a geometric invariant of a curve analytically?

(ii) How does the very definition of "curve" give analytic content to what is geometrically an intuitive idea?

(iii) Why are curves one-dimensional objects?

(iv) What does it mean for a geometric property or invariant of a curve to be independent of parametrization?

(v) What does it mean for two curves to be congruent?

(vi) What does it mean to say that the three geometric invariants of arclength, curvature, and torsion together characterize the shape, but perhaps not the position, of a curve?

You should include any other points you feel will help convey that you understand the geometric nature of curves and that that nature can be revealed analytically via functions from vector calculus.

13. In this problem we will be concerned with finding curves having prescribed curvature and torsion properties. Consider a curve **r** of the form

$$\mathbf{r}(t) = \int f(t)\cos t \, dt \, \mathbf{i} + \int f(t)\sin t \, dt \, \mathbf{j} + \int f(t)g(t) \, dt \, \mathbf{k},$$

where f and g are smooth functions and f is positive.

(a) Compute the curvature and torsion of **r** in terms of these functions and their derivatives.

(b) Use those formulas to construct a specific curve $\mathbf{r}_1$ which has constant curvature, but nonconstant torsion.

(c) Use the formulas to construct a specific curve $\mathbf{r}_2$ which has constant torsion, but nonconstant curvature. (Hint: Select $g(t) = \sinh t$ in parts (b) and (c). Then choose f appropriately.)

(d) Have *Mathematica* attempt to carry out the symbolic antidifferentiations in the components of the curves $\mathbf{r}_1$ and $\mathbf{r}_2$ with the functions you have chosen. If you succeed, proceed to part (e). If you are unsuccessful, then do the integrals numerically (i.e., with `NIntegrate`) to find the antiderivatives.

(e) Use your results from (d) to plot the two curves.

14. Fix a circle of radius a. Let a second circle of radius b roll around the outside of the first circle without slipping. The *epicycloid* is the curve traced by a point on the circumference of the rolling circle.

    (a) Find parametric equations for the epicycloid.

    (b) Graph the epicycloid for at least three values of a and b.

    (c) Compute the speed and curvature of the epicycloid. Relate any points where these invariants are zero or undefined to geometric features of the graph. (Hint: The standard program in Chapter 3 may take a long time to compute the curvature, since it gets hung up trying to `Simplify` intermediate steps. Try computing from the formulas for variable-speed curves, stripping out all the simplification commands until you get to the end. You get a simpler result if you first compute the square of the curvature and then take the square root.)

    (d) Find the evolute of the epicycloid. When $a = 3$ and $b = 1$, graph the epicycloid, the fixed circle, and the evolute on the same axes. (The hint for (c) applies here as well.)

15. The *hypocycloid* is the path traced by a point on the circumference of a wheel of radius b rolling on the inside of another circle of radius a. Follow the instructions for Problem 14, using the hypocycloid instead of the epicycloid. Use $a = 11$, $b = 7$ when you graph the hypocycloid and its evolute.

16. A reflector is attached to a bicycle wheel of radius a at a distance b from the center. As the wheel rolls without slipping, the reflector sweeps out a curve called a *trochoid*. Follow the instructions for Problem 14, using the trochoid instead of the epicycloid, and replacing the fixed circle in part (d) by the straight line on which the wheel is rolling.

17. The *Mathematica* programs given in Chapter 3 for computing the Frenet frame, curvature, and torsion violate good programming practice in at least two ways.

    (a) The given programs assume that every parametric curve is defined in terms of the same parameter t. As they stand, they could not be used with the reparametrization of the helix, where we rewrote the curve in terms of a new variable s. Write new versions of the programs that allow you to specify the name of the variable as one of the arguments to the program. In particular, you should be able to type UT[unithelix, s] to compute the unit tangent vector to the reparametrized helix. To verify that your programs are correct, apply them to the reparametrized helix, and compare the answers with the ones given in the text.

    (b) The given programs are inefficient (and slow), especially if you need to compute all the invariants for a single curve. For example, when you compute the torsion, the program must compute the binormal, which causes it to compute the unit tangent and principal normal, etc. Write new versions of the programs (possibly taking different arguments) that avoid recomputing things.

    (c) Finally, write a program called **geometricInvariants** that takes two arguments: a parametrization of a curve and the name of the parameter. This program should return a list containing the components of the Frenet frame along with the speed, curvature, and torsion. (You should probably read about the **Module** command before attempting this problem.)

Glossary of Some Useful *Mathematica* Objects

Commands

Clear Clears the definition of a variable.

Cross Cross product command in *Mathematica* 3.0.

CrossProduct Cross product command in the **Calculus `VectorAnalysis`** package.

D Differentiates an expression.

DSolve Solves a differential equation analytically.

Evaluate Forces evaluation of an expression; needed in some plotting commands.

Graphics Creates a primitive graphics object such as a **Disk** or a **Line**.

ImplicitPlot Plots a curve defined implicitly.

Integrate Integrates an expression.

`Module` Programming construct; used to write programs with local variables.

`NDSolve` Solves a differential equation numerically.

`NIntegrate` Integrates an expression numerically.

`ParametricPlot` Plots curves in the plane defined by parametric equations.

`ParametricPlot3D` Plots curves or surfaces in space defined by parametric equations.

`Text` Used with `Show` and `Graphics` to add text to a graph.

Built-in Functions

`ArcCosh` Inverse hyperbolic cosine function.

`ArcSinh` Inverse hyperbolic sine function.

`Sech` Hyperbolic secant function.

Options and Directives

`Boxed` Determines if a box is drawn around a three-dimensional graphic.

`GrayLevel` Determines the shade of gray in which to plot a curve, with 0 representing black and 1 representing white.

`PlotStyle` Specifies the style of curves to be plotted.

`Ticks` Controls the placement of tick marks along axes in a graph.

`ViewPoint` Determines the position from which a three-dimensional graph is viewed.

Packages

`Calculus`VectorAnalysis`` Needed for `CrossProduct`.

`Graphics`ImplicitPlot`` Needed for `ImplicitPlot`.

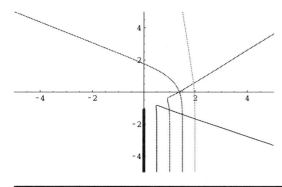

Chapter 4

KINEMATICS

The study of curves in space is of interest not only as a topic in geometry, but also for its application to the motion of physical objects. In this chapter, we develop a few topics in mechanics from the point of view of the theory of curves. Additional applications to physics will be considered in Chapter 9, *Physical Applications of Vector Calculus*.

Newton's Laws of Motion

In the last chapter, we alluded to the fact that a parametrized curve $t \mapsto \mathbf{r}(t) = (x(t), y(t), z(t))$, with t representing time and $\mathbf{r}$ representing position, may be used to describe the motion of an object in space. This works provided the object is small enough compared to the distances over which it travels that we may represent it as a point. Even if the moving object is large, it is often convenient to replace it by an idealized particle concentrated at its center of mass; it often turns out that the center of mass of the macroscopic object moves in the same way as this idealized particle. We will, therefore, only consider point objects in this chapter.

The basic principles of both motion and the calculus were enunciated by Isaac Newton in his book *Philosophiæ Naturalis Principia Mathematica*, or *Mathematical Principles of Natural Philosophy*, in 1687. The first two of Newton's laws of motion read as follows [I. Newton, *The Principia*, translated by A. Motte, republished in the Great Minds Series, Prometheus Books, 1995, p. 19]:

Every body perseveres in its state of rest, or of uniform motion in a right line, unless it is compelled to change that state by forces impressed thereon.

The alteration of motion is ever proportional to the motive force impressed; and is made in the direction of the right line in which that force is impressed.

In deference to Newton, it is traditional in mechanics to use his *dot notation* for derivatives with respect to time. (This is the only place in calculus where Newton's notation has triumphed over that of Leibnitz.) For an object represented mathematically by a parametrized curve $t \mapsto \mathbf{r}(t)$, the first law says that the velocity $\mathbf{v}(t) = \dot{\mathbf{r}}(t)$ remains constant in the absence of outside forces. The second law says that the rate of change of the velocity, or in other words the acceleration $\mathbf{a}(t) = \dot{\mathbf{v}}(t) = \ddot{\mathbf{r}}(t)$, is proportional to the force exerted. We usually summarize both laws in a single equation: $\mathbf{F} = m\mathbf{a}$, where $\mathbf{F}$ is the total force acting on the object, $\mathbf{a}$ is the acceleration, and m is the mass of the object. (Newton established later in his book that the mass is the correct constant of proportionality.) This law is a second-order differential equation for the vector-valued function $\mathbf{r}(t)$, or a system of three second-order scalar differential equations for the components $x(t)$, $y(t)$, and $z(t)$ of $\mathbf{r}(t)$.

A number of special cases of the equation $\mathbf{F} = m\ddot{\mathbf{r}}$ are of particular interest. In studying them, it is useful to introduce the *kinetic energy*, $E_{\text{kin}} = \frac{1}{2}m\|\mathbf{v}\|^2$. We find that

$$\frac{d}{dt}E_{\text{kin}} = \frac{d}{dt}\left(\tfrac{1}{2}m\|\mathbf{v}\|^2\right) = \tfrac{1}{2}m\frac{d}{dt}(\mathbf{v}\cdot\mathbf{v}) = \mathbf{v}\cdot(m\dot{\mathbf{v}}) = \mathbf{v}\cdot\mathbf{F}. \tag{1}$$

Thus, if $\mathbf{F}$ is always perpendicular to the velocity $\mathbf{v} = \dot{\mathbf{r}}$, then (1) says that the kinetic energy remains constant, and that the object moves with constant speed. This case arises when a charged particle moves under the influence of a magnetic field $\mathbf{B}$. The magnetic force on a charged particle with charge q and position vector $\mathbf{r}$ is

$$\mathbf{F}_{\text{mag}} = \frac{-q}{c}\mathbf{B}\times\dot{\mathbf{r}}, \tag{2}$$

where c is a universal constant, the speed of light in a vacuum. Since the cross product $\mathbf{B}\times\dot{\mathbf{r}}$ is perpendicular to $\dot{\mathbf{r}}$, the charged particle moves at constant speed v. From the theory of constant-speed curves in the last chapter, we can compute the curvature of the motion:

$$\kappa = \frac{1}{v^2}\|\mathbf{a}\| = \frac{q}{mv^2c}\|\mathbf{B}\times\dot{\mathbf{r}}\| = \frac{q}{mvc}\|\mathbf{B}\|\sin\theta,$$

where θ is the angle between the magnetic field $\mathbf{B}$ and the velocity $\mathbf{v}$.

Another case of interest is that of motion in a *central force field*, where the force $\mathbf{F}$ on the object is a *time-independent* function of the position $\mathbf{r}$, and this function is of the very special form

$$\mathbf{F} = f(r)(\text{unit vector pointing toward the origin}),$$

where $r = \|\mathbf{r}\|$ is the distance to the origin, and f is an ordinary scalar function. Since the unit vector pointing toward the origin is $-\mathbf{r}/r$, motion in a central force field is given by the differential equation

$$\ddot{\mathbf{r}} = -\frac{f(r)}{mr}\mathbf{r} = g(r)\mathbf{r}. \tag{3}$$

This differential equation is *autonomous*, i.e., the right-hand side of (3) does not depend explicitly on t. When $f(r) > 0$, the force on the object points toward the origin with magnitude $f(r)$. This is the case of an *attractive force*. The standard example, as originally studied by Newton, is a planet moving around a sun (located at the origin) under the influence of gravity. In this case, the *inverse square law* says the magnitude $f(r)$ of the force is inversely proportional to r^2, the square of the distance to the origin. More precisely, *Newton's law of gravitation* says that $f(r) = GMm/r^2$, where M is the mass of the Sun, m is the mass of the planet, and G is a universal gravitational constant. Electrical forces between point charges of opposite sign are also attractive and also obey the inverse square law, but with $f(r)$ now depending on the product of the charges. When $f(r) < 0$, the force on the object points away from the origin, and this is the case of a *repulsive force*. For example, the electric force between charges of the same sign is repulsive.

To study equation (3) for motion in a central force field, it is traditional to introduce the *angular momentum* vector

$$\mathbf{L} = m\mathbf{r} \times \mathbf{v}. \tag{4}$$

Differentiating (4), and using the product rule, we find that

$$\dot{\mathbf{L}} = m\dot{\mathbf{r}} \times \mathbf{v} + m\mathbf{r} \times \dot{\mathbf{v}} = m\mathbf{v} \times \mathbf{v} + m\mathbf{r} \times g(r)\mathbf{r} = 0 + 0 = 0.$$

Thus $\mathbf{L}$ does not change with time; it is a fixed quantity. This is the *law of conservation of angular momentum*. Since $\mathbf{L}$ is perpendicular to both $\mathbf{r}$ and $\mathbf{v}$, if the latter are not collinear, then $\mathbf{L}$ is normal to the unique plane containing them. Constancy of $\mathbf{L}$ therefore implies that the plane containing $\mathbf{r}$ and $\mathbf{v}$ doesn't change, and thus *the curve traced out by the object lies in a plane*.

We may obtain the same conclusion using the theory of the torsion of curves from the last chapter. There we had the formula

$$\tau = \frac{(\dot{\mathbf{r}} \times \ddot{\mathbf{r}}) \cdot (\ddot{\mathbf{r}})^{\cdot}}{\|\dot{\mathbf{r}} \times \ddot{\mathbf{r}}\|^2}$$

for the torsion of the curve traced out by $\mathbf{r}(t)$. Substituting $g(r)\mathbf{r}$ for $\ddot{\mathbf{r}}$, we obtain the equation

$$\tau = \frac{(\dot{\mathbf{r}} \times g(r)\mathbf{r}) \cdot (g(r)\mathbf{r})^{\cdot}}{\|\dot{\mathbf{r}} \times g(r)\mathbf{r}\|^2}.$$

But

$$(g(r)\mathbf{r})^{\cdot} = (g(r))^{\cdot}\,\mathbf{r} + g(r)\dot{\mathbf{r}},$$

which is a sum of a scalar multiple of **r** and a scalar multiple of $\dot{\mathbf{r}}$, hence perpendicular to $\dot{\mathbf{r}} \times g(r)\mathbf{r}$. Thus the numerator vanishes and $\tau = 0$; i.e., the motion is planar. The experimental evidence for this conclusion is exceptionally good. In our own solar system, it turns out that to a high degree of approximation, the motions of the Sun and of all of the planets except Pluto lie in a single plane called the *ecliptic*. The motion of Pluto also lies in a plane, though it is slightly tilted with respect to the ecliptic.

Kepler's Laws of Planetary Motion

By analyzing equation (3) using his theory of gravitation, Newton [*Principia*, Book III] was able to derive all the laws of planetary motion previously found purely phenomenologically by Kepler. One of these laws is that the motion of a planet stays in a plane; as we have seen, this is a property of motion in any central force field. Another law of a similar nature is that a planet "sweeps out equal areas in equal lengths of time." In other words, as a planet moves from a position P to a position Q around the sun at position S, the area of the region $\bigtriangledown SPQ$ bounded by the line segments $\overline{SP}$ and $\overline{SQ}$ and by the path (or orbit) of the planet from P to Q is proportional to the time taken in going from P to Q and otherwise independent of the choices of P and Q. To derive this law, note that for P and Q very close together corresponding to times t_0 and $t_0 + \Delta t$, $\bigtriangledown SPQ$ is approximately a triangle, so its area is approximately

$$\frac{1}{2}\|\mathbf{r}(t_0) \times \mathbf{r}(t_0 + \Delta t)\| \approx \frac{1}{2}\|\mathbf{r}(t_0) \times (\mathbf{r}(t_0) + (\Delta t)\dot{\mathbf{r}}(t_0))\| = \frac{1}{2}\Delta t \,\|(\mathbf{r} \times \dot{\mathbf{r}})(t_0)\|.$$

But $\mathbf{r} \times \dot{\mathbf{r}}$ is just $\mathbf{L}/m$, which is constant, so the area is the constant $\|\mathbf{L}\|/2m$ times the time interval Δt. The following diagram shows the region $\bigtriangledown SPQ$ along with a second region of equal area:

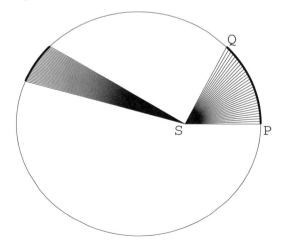

Kepler's other two laws depend on the fact that gravitation obeys an inverse square law. They say that planets travel in ellipses or circles, and that more generally, objects moving under the central force field of the sun move in conic sections: circles, ellipses, hyperbolas, or parabolas, with the sun at one focus. Furthermore, for a planet in an elliptical orbit, the period of the orbit (the time it takes for one complete revolution around the sun) is proportional to the $\frac{3}{2}$ power of the major axis of the orbit. These two laws will be studied in Problem Set D.

Studying Equations of Motion with *Mathematica*

Mathematica provides a convenient framework for studying Newton's equation of motion $\mathbf{F} = m\ddot{\mathbf{r}}$, once we understand how to use the two *Mathematica* commands for solving differential equations, DSolve and NDSolve. DSolve works only with the (somewhat limited) class of differential equations that can be solved explicitly in closed form. It gives its output as a list of substitution rules for the unknown functions, possibly involving various parameters. To illustrate how it works, we solve the equation of motion for a charged particle with charge q, mass m, and initial velocity $\mathbf{v}(0) = (v_1, v_2, v_3)$, in a uniform magnetic field $\mathbf{B} = b\mathbf{k}$. We take the initial position to be at the origin.

```
In[1]:=  <<Calculus`VectorAnalysis`
```

```
In[2]:=  rhs = (-q*b/(c*m))*
            CrossProduct[{0, 0, 1},
            {x'[t], y'[t], z'[t]}]
```

$$Out[2]= \left\{ \frac{bqy'[t]}{cm}, -\frac{bqx'[t]}{cm}, 0 \right\}$$

```
In[3]:=  sol = DSolve[{x''[t] == rhs[[1]],
            y''[t] == rhs[[2]],
            z''[t] == 0,
            x'[0] == v1,
            y'[0] == v2,
            z'[0] == v3,
            x[0] == 0,
            y[0] == 0,
            z[0] == 0},
            {x[t], y[t], z[t]}, t]
```

Out[3]=
$$\left\{\left\{x[t] \rightarrow \frac{cmv2 - cmv2\,Cos\left[\frac{bqt}{cm}\right] + cmv1\,Sin\left[\frac{bqt}{cm}\right]}{bq}\right.\right.,$$

$$y[t] \rightarrow \frac{-cmv1 + cmv1\,Cos\left[\frac{bqt}{cm}\right] + cmv2\,Sin\left[\frac{bqt}{cm}\right]}{bq},$$

$$\left.\left. z[t] \rightarrow tv3\right\}\right\}$$

To use this answer, we need to extract the solution with `First`, and make appropriate substitutions with the replacement operator. For example, if we take $v1 = 0$ and set b, q, c, and m all to 1, we obtain:

```
In[4]:=  {x[t], y[t], z[t]} /. First[sol] /.
         {v1->0, b->1, q->1, c->1, m->1} // Simplify
```

Out[4]= $\{v2 - v2\,Cos[t], v2\,Sin[t], tv3\}$

We can recognize this from the last chapter as the parametrization of a helix.

The `NDSolve` command works slightly differently. It can deal with essentially any system of ordinary differential equations, but only produces a numerical approximation to a solution, in the form of substitution rules involving an `InterpolatingFunction`. Since `NDSolve` works numerically, all parameters and initial conditions in the equations must be specified explicitly, and we must specify a domain for the independent variable. For example, if we wanted to use `NDSolve` in the above example, we would have had to specify all the constants at the beginning. For instance, if we take $v1 = 0$, $v2 = v3 = 1$ and set b, q, c, and m all to 1, we obtain:

```
In[5]:=  rhsNumerical = rhs /. {b->1, q->1, c->1, m->1}
```

Out[5]= $\{y'[t], -x'[t], 0\}$

```
In[6]:=  solNumerical = NDSolve[
         {x''[t] == rhsNumerical[[1]],
         y''[t] == rhsNumerical[[2]],
         z''[t] == 0,
         x'[0] == 0,
         y'[0] == 1,
         z'[0] == 1,
         x[0] == 0,
         y[0] == 0,
         z[0] == 0},
         {x, y, z}, {t, 0, 6Pi}]
```

Out[6]= $\{\{x \rightarrow$ InterpolatingFunction$[\{\{0., 18.8496\}\}, <>],$
$y \rightarrow$ InterpolatingFunction$[\{\{0., 18.8496\}\}, <>],$
$z \rightarrow$ InterpolatingFunction$[\{\{0., 18.8496\}\}, <>]\}\}$

```
In[7]:= ParametricPlot3D[Evaluate[
          {x[t], y[t], z[t]} /. First[solNumerical]],
          {t, 0, 6Pi}, AspectRatio -> 2];
```

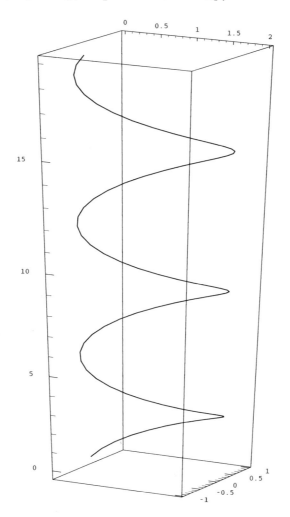

We indeed see the characteristic shape of a helix.

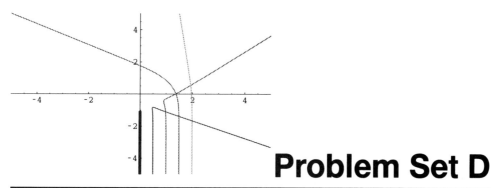

Problem Set D

KINEMATICS

1. This problem continues the study, begun in Chapter 4, of the motion of a charge q under the influence of a magnetic field $\mathbf{B}(\mathbf{r})$. We assume $\mathbf{B}(\mathbf{r})$ varies with position in space but is *static*, i.e., time-independent, and that there are no other forces acting on the charge. For convenience in the calculations, take $q = m = c = 1$ (i.e., ignore constant factors).

   (a) Using DSolve, redo the exact solution of the equations of motion in the case where $\mathbf{B} = \mathbf{k}$ is constant and $\mathbf{v}(0)$, the initial value of $\mathbf{v}$ at time $t = 0$, is given but arbitrary. (The *Mathematica* program from Chapter 4 can be found in a Notebook on the disk that accompanies this book.) Show that the charge moves in a circle if $\mathbf{v}(0)$ is a nonzero vector with vanishing k-component, and that it moves in a helix if $\mathbf{v}(0)$ has both a nonzero i- or j-component and a non-zero k-component. Compute the curvature and torsion to verify the condition for a right circular helix. By choosing specific initial data, draw pictures of both a circular and a helical trajectory.

   (b) Solve the equations of motion numerically, using NDSolve, in the case where $\mathbf{B}(x, y, z) = (-y\mathbf{i} + x\mathbf{j})/(x^2 + y^2)$. This corresponds to the magnetic field due to a current flowing along the z-axis. Using ParametricPlot3D, graph the resulting trajectories. Try to interpret the results. Use the following initial conditions:

(i) $\mathbf{r}(0) = (1, 0, 0)$, $\mathbf{r}'(0) = (1, 0, 0)$.

(ii) $\mathbf{r}(0) = (1, 0, 0)$, $\mathbf{r}'(0) = (2, 0, 0)$.

(iii) $\mathbf{r}(0) = (1, 0, 0)$, $\mathbf{r}'(0) = (1, 0, 1)$.

(iv) $\mathbf{r}(0) = (1, 0, 0)$, $\mathbf{r}'(0) = (2, 0, 1)$.

2. One of the simplest cases where Newton's law of gravitation can be applied is that of the motion of a ball near the surface of the Earth.

(a) We begin by neglecting air resistance. Let's choose coordinates so that the y-axis points straight up and the x-axis is along the ground. Then the gravitational force on the ball is $\mathbf{F} = (0, -mg)$, where g is the acceleration due to gravity. Suppose a batter hits a baseball from the position $(0, h)$ with initial velocity $\mathbf{v}_0 = (u_1, u_2)$ at time $t = 0$, and thereafter the only force acting on the ball is that of gravity. Solve the equations of motion (with DSolve) for the trajectory $\mathbf{r}(t)$ of the ball, and show that the ball moves in an arc of a parabola (assuming $u_1 > 0$). Find the distance d traveled by the ball before it hits the ground, and the time t_0 taken to travel that distance. (You need to solve the vector equation $\mathbf{r}(t_0) = (d, 0)$, with a *positive* value of t_0. Your answers should involve h, u_1, u_2, and g.)

(b) Suppose that the speed $v_0 = \sqrt{u_1^2 + u_2^2}$ at which the ball leaves the bat is fixed, but that the batter can adjust the angle θ at which the ball is hit. In other words, take $\mathbf{v}_0 = v_0(\cos\theta, \sin\theta)$, where $-\pi/2 \le \theta \le \pi/2$. (Here $\theta = -\pi/2$ corresponds to the batter hitting a foul tip that bounces off home plate, and $\theta = \pi/2$ corresponds to the batter popping the ball straight up.) Find the value of θ that maximizes the distance d traveled by the ball. Also find that value of θ that maximizes the time t_0 that the ball remains in the air. (Your answers should involve h, v_0, and g.) Do your answers match your expectations? (Keep in mind that h is very small compared to the maximum height of a fly ball.)

(c) Assuming the batter adjusts θ to maximize the distance d, obtain numerical values for d (in feet) and t_0 (in seconds) if $g = 32\,\text{ft/sec}^2$, $h = 4\,\text{ft}$, and $v_0 = 100\,\text{ft/sec}$ (about $68\,\text{mi/hr}$). Do your values for d and t_0 seem reasonable?

(d) Suppose the center field fence is 400 feet from home plate and 10 feet high. How much harder does the batter have to hit the ball so that it reaches the fence?

(e) How fast does the batter have to hit the ball so that it *clears* the center field fence? (This problem is harder; it requires redoing the calculations to maximize, as a function of θ, the distance at which the height of the ball is 10 feet instead of 0 feet, and then solving to see for what value of v_0 this distance reaches 400 feet.)

(f) To obtain a more reasonable model, we should include the effect of air resistance. Then the total force on the ball is

$$\mathbf{F} = (0, -mg) - \frac{mg}{v_{\text{term}}}\dot{\mathbf{r}},$$

where the air resistance term involves the quantity v_{term}, the *terminal velocity* of the ball. (This is the speed at which the ball would eventually fall if dropped from a great height.) Redo the solution of the equations of motion. Plot the trajectory of the ball (with the initial conditions from part (c)), both with and without air resistance (assuming, say, that $v_{\text{term}} = 300\,\text{ft/sec}$). Superimpose the two plots. Do you see any significant difference?

3. Consider an object of mass $m = 1$ with position vector $\mathbf{r}(t) = (x(t),\, y(t),\, z(t))$ moving in a central force field

$$\mathbf{F} = -\frac{1}{\|\mathbf{r}\|^3}\mathbf{r},$$

which is an idealized model of a planet moving around a massive sun. (For convenience, we've chosen units so as to set the attractive constant GM to 1.) For simplicity, we'll assume the object moves in the x-y plane.

(a) If the planet moves in a circular orbit, then its speed v is a constant. Why? (Use equation (1) in Chapter 4.)

(b) Continue to assume that the planet moves in a circular orbit. What is the relationship between the radius r of the orbit and the speed v of the planet? Also find the period T of the orbit as a function of r. Your answer should be a form of one of *Kepler's laws of planetary motion*. (Hint: In this case r is a constant, so the motion is described by the angular coordinate θ. The speed is $v = r\omega$, where $\omega = \dot{\theta}$ is the angular velocity, a constant of the motion, and $\mathbf{r} = (r\cos\theta,\, r\sin\theta) = (r\cos\omega t,\, r\sin\omega t)$. Differentiate twice with respect to t and substitute in Newton's law of gravitation.)

(c) Use **NDSolve** to solve numerically the equations of motion:

$$\ddot{x}(t) = -\frac{x(t)}{(x(t)^2 + y(t)^2)^{\frac{3}{2}}}, \qquad \ddot{y}(t) = -\frac{y(t)}{(x(t)^2 + y(t)^2)^{\frac{3}{2}}},$$

with the initial conditions $x(0) = 1$, $y(0) = 0$, for $0 \le t \le 10$, in the three cases:

(i) $\dot{x}(0) = 0$, $\dot{y}(0) = 1$.

(ii) $\dot{x}(0) = 0$, $\dot{y}(0) = 1.2$.

(iii) $\dot{x}(0) = 0$, $\dot{y}(0) = 2$.

Display the trajectory in each case using **ParametricPlot**. (You may need to adjust the **AspectRatio** so that the scales on the two axes are the same.)

Which conic sections do you observe? In one case, you may want to allow a longer time interval to clearly identify the trajectory. In another case, you may want to augment the time interval to allow "negative time."

4. This problem is similar to Problem 3, except that the central force field doesn't quite obey the inverse square law. Suppose now that a particle of mass $m = 1$ with position vector $\mathbf{r}(t) = (x(t),\, y(t),\, z(t))$ is moving in a central force field

$$\mathbf{F} = -\frac{1}{\|\mathbf{r}\|^{\alpha}}\mathbf{r},$$

where the parameter α is not necessarily equal to 3.

(a) As before, if the planet moves in a circular orbit, find the relationship between the radius r of the orbit and the speed v of the planet. Your answer should depend on α. Also find the period T of the orbit as a function of r and α. Explain how you can use your answer to provide an experimental test of the inverse square law from the fact that the radius of the orbit of Mars is 1.524 times that of the Earth, while a year on Mars is 687.0 days long (compared with 365.26 days on Earth). Keep in mind that you cannot expect perfect accuracy since we have neglected the eccentricities of the orbits and the fact that the planets exert gravitational forces on one another.

(b) Repeat the calculations of Problem 3(c), but with $\alpha = 4$ (an inverse cube law) instead of $\alpha = 3$. How do the trajectories differ from what they would be with an inverse square law, and why?

5. This problem explores two other central force fields that arise in molecular and nuclear physics.

(a) The Yukawa force field, which models short-range attractive forces such as those that hold an atomic nucleus together, gives the equation of motion

$$\ddot{\mathbf{r}} = \frac{\mathbf{r}}{mr}\frac{d}{dr}\left(b\frac{e^{-ar}}{ar}\right),$$

where a and b are positive constants. (The quantity b measures the strength of the force field, and a measures its "range." More precisely, the force is significant only at distances on the order of $1/a$. Of course we are using a somewhat inappropriate model, since nuclear forces can only be understood using quantum mechanics, and we are using classical mechanics here.)

Find, for this force field, the relationship between the speed v and the radius r for circular orbits. Compute the period T as a function of r. (The result will depend on a and b.) Compare the results to the situation in Kepler's laws, where v is inversely proportional to $\sqrt{r}$ and T is proportional to $r^{3/2}$. To visualize what happens, it is convenient to use **LogLogPlot**, in the package

`Graphics'Graphics'`, to draw a log-log plot of T as a function of r. (For convenience, take $m = a = b = 1$.) Recall that the log-log plot of a power function is a straight line, with slope equal to the exponent of the power law. So the deviation of the log-log plot from a straight line measures the deviation from a power law.

(b) The van der Waals force field, which models the interaction between atoms or molecules, is mildly attractive at larger distances, and strongly repulsive at very short distances. (This reflects the fact that molecules resist being pushed into one another. Again, we are using a somewhat inappropriate model, since molecular forces ought to be modeled with quantum mechanics.) Try the force field

$$\ddot{\mathbf{r}} = \frac{\mathbf{r}}{r}\frac{d}{dr}\left(\frac{1}{r^6} - \frac{1}{r^{12}}\right),$$

which is a version of a commonly used model. Plot the force as a function of r (with a suitable `PlotRange`) to see that the force is indeed strongly repulsive for $r < 1$ and mildly attractive for some bigger values of r, especially for $1.2 < r < 2$. The correct function to plot is

$$\frac{d}{dr}\left(\frac{1}{r^6} - \frac{1}{r^{12}}\right).$$

Use `NDSolve` to compute the trajectories for objects starting at $y = -5$ and at $x = 0, 0.5, 1.0, 1.5, 2.0$, in each case with initial velocity $(0, 1)$. You should observe what physicists call *scattering* in various directions. How can you explain the results?

6. Celestial mechanics is the study of the motion of celestial objects under the influence of gravity. The study of the motion of two objects under Newtonian gravitation is known as the *2-body problem* in celestial mechanics. The 2-body problem can be solved completely. The *3-body problem*, on the other hand, cannot be solved completely and sometimes exhibits strange behavior. We will investigate some simple 3-body systems in this and the next problem.

Assume we have a "solar system" with a massive fixed Sun at the origin, containing a planet and a comet that interact with each other as well as with the Sun. For simplicity, we assume that the mass of the planet is negligible with respect to the mass of the sun, and the mass of the comet is negligible with respect to that of the planet. Thus, the planet exerts a force on the comet but not vice versa.

(a) Represent the coordinates of the planet by $(x_1(t), y_1(t))$ and those of the comet by $(x_2(t), y_2(t))$. As in Problem 3(c), use `NDSolve` to solve numerically the equations of motion

$$\ddot{x}_j(t) = -\frac{x_j(t)}{\left(x_j(t)^2 + y_j(t)^2\right)^{\frac{3}{2}}}, \quad \ddot{y}_j(t) = -\frac{y_j(t)}{\left(x_j(t)^2 + y_j(t)^2\right)^{\frac{3}{2}}}, \quad j = 1, 2,$$

with the initial conditions $x_1(0) = 1$, $x_2(0) = 2$, $\dot{x}_1(0) = 0$, $\dot{x}_2(0) = 0$, $y_1(0) = 0$, $y_2(0) = 0$, $\dot{y}_1(0) = 1$, $\dot{y}_2(0) = 0.1$, for $0 \leq t \leq 15$. These equations describe the motion when the planet and the comet move independently (i.e., without interacting), in the sun's central force field. Plot the resulting trajectories (as in 3(c)) on a single set of axes. You should find that the planet moves in a circular orbit, and that the comet moves in an elongated elliptical orbit that comes very close to the sun. By adjusting `PlotRange`, graph the comet's orbit near the sun. Does it crash? If your trajectory is jagged, try increasing `PlotPoints`.

(b) Now add the effect of the planet's gravitation on the comet and redo the calculation (for t running from 0 to 30) and plot as in part (a). Assume for simplicity that the mass of the planet is one-tenth that of the sun, so that the equation for $\mathbf{r}_2 = (x_2, y_2)$ now becomes

$$\ddot{\mathbf{r}}_2 = -\frac{\mathbf{r}_2}{\|\mathbf{r}_2\|^3} - 0.1\frac{\mathbf{r}_2 - \mathbf{r}_1}{\|\mathbf{r}_2 - \mathbf{r}_1\|^3}.$$

Explain why this is the correct equation of motion. What is the effect of the planet on the motion of the comet? (Since the equations are more complicated, don't be surprised if *Mathematica* takes longer to solve them.)

7. In this problem we investigate another 3-body problem, that of a planet moving around a *double* star.

(a) Suppose that two suns of equal mass M rotate in a circle under the influence of each other's gravity. (This is a reasonable classical model for a double star.) Let $\mathbf{r}_j = (x_j, y_j)$, $j = 1, 2$, denote the position of the jth star. If the stars rotate opposite each other in a circle of radius R at fixed angular velocity ω, then

$$\mathbf{r}_1(t) = (R\cos\omega t, R\sin\omega t), \quad \mathbf{r}_2(t) = -\mathbf{r}_1(t) = (-R\cos\omega t, -R\sin\omega t).$$

Since the distance between the stars is $2R$, the equations of motion for Newtonian gravitation are then given by

$$M\ddot{\mathbf{r}}_j = -GM^2\frac{\mathbf{r}_j - (-\mathbf{r}_j)}{(2R)^3},$$

or

$$\ddot{\mathbf{r}}_j = -GM\frac{\mathbf{r}_j}{4R^3}.$$

Differentiate $\mathbf{r}_1$ twice and find the relationship between M, R, and ω.

(b) Suppose that we've normalized things so that $R = 0.1$ and $\omega = 1$. What does this tell you about GM? What is the equation of motion for a planet moving about this double star?

(c) From the point of view of a planet far away from the double star, things should not be that different from the case of a planet moving around a single star of mass $2M$. As in Problem 3(c), solve and plot the trajectory for a planet moving around a single star of mass $2M$ in a circular orbit in the x-y plane, centered at the origin and with radius 1. Then solve and plot the trajectory for the same planet moving with the same initial conditions around the double star in (b) above. The latter looks circular, but the trajectory is a little ragged. That's because there are *wobbles* in the orbit. Plot separately (using `Plot`) the differences between the x- and y-coordinates for the two trajectories to see the magnitude of the wobble.

Glossary of Some Useful *Mathematica* Objects

Commands

`DSolve` Symbolic differential equation solver.

`First` Gets the first element of a list.

`LogLogPlot` Draws a log-log plot.

`NDSolve` Numerical differential equation solver.

`ParametricPlot` Plots a curve parametrically in 2-space.

`ParametricPlot3D` Plots a curve parametrically in 3-space.

`PolarPlot` Plots a curve in polar coordinates.

Options and Directives

`Dashing` Used to draw dashed lines.

`PlotRange` Specifies the range displayed in a plot.

`PlotStyle` Specifies style for `Plot` and `ParametricPlot`.

`PointSize` Specifies (relative) point size.

`RGBColor` Specifies a color. Use `RGBColor[1,0,0]` for red; `RGBColor[0,1,0]` for green; `RGBColor[0,0,1]` for blue.

`Thickness` Specifies (relative) line thickness.

Packages

`Graphics`Graphics`` needed for `PolarPlot`, `LogLogPlot`.

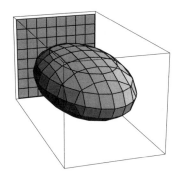

Chapter 5

DIRECTIONAL DERIVATIVES

In the last two chapters, we studied curves in space, that is, *vector-valued functions of a single scalar variable*. We move on now to functions of two variables, or equivalently, *scalar-valued functions of a single vector variable* $\mathbf{x} = (x, y)$. Such a function f is a rule, usually given by a formula, that associates a value $f(\mathbf{x}) = f(x, y)$ to any $\mathbf{x} = (x, y)$ in a subset of $\mathbf{R}^2$ called the *domain* of the function. We are interested primarily in the case where f is continuous and differentiable. (Sometimes this may involve cutting down the domain. For example, if

$$f(x, y) = \sqrt{1 - x^2 - y^2},$$

then the natural domain where f is defined is $\{(x, y) : x^2 + y^2 \leq 1\}$, but it's convenient to restrict to the smaller domain $\{(x, y) : x^2 + y^2 < 1\}$. Otherwise, f is not differentiable at the "boundary points" where $x^2 + y^2 = 1$.) Our goals are to learn how to visualize the function, how to differentiate it, and how to use the visualizations and derivatives to understand algebraic and geometric properties.

Visualizing Functions of Two Variables

There are two standard ways to visualize a function f of two variables: looking at its *graph*, that is, the set of points

$$\{(x, y, z) : z = f(x, y)\}$$

in $\mathbf{R}^3$; and looking at its *level curves*, which are the curves in $\mathbf{R}^2$ given implicitly by the equations $f(x, y) = c$ with c a constant. One of the main points of this chapter is to discuss what information about f is encoded in the graph and in the level curves. While sketching either the graph or the level curves of f is often hard to do by hand, both are easy to display with *Mathematica*.

Three-Dimensional Graphs

`Plot3D` is the basic *Mathematica* routine for graphing a function of two variables. Let's start with a simple example, the function

$$f(x, y) = y \left(1 - \frac{1}{(x^2 + y^2)} \right), \quad (x, y) \neq (0, 0).$$

It has an interesting connection with fluid flow, which we'll discuss later. We encode the function as a *Mathematica* expression and then plot it.

```
In[1]:=  flowFunction = y*(1 - 1/(x^2 + y^2));
```

```
In[2]:=  Plot3D[flowFunction, [x, -2, 2], [y, -2, 2]];
```

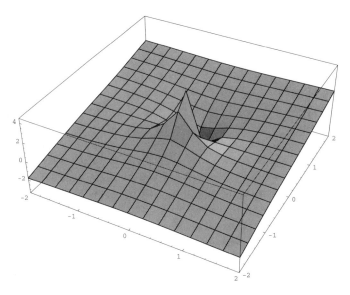

This is not the greatest picture, but it does tell us some things about the function f. First of all, the graph is highest where f is the largest, has peaks where f has values that are higher than in the surrounding regions, and has troughs where f has values that are lower than in the surrounding regions. We can see a small region near the origin along the negative y-axis where f gets large, and a small region near the origin along the positive y-axis where f gets to be quite negative. (Recall from Chapter 2 that *Mathematica*'s default `ViewPoint` in three-dimensional graphics is $(1.3, -2.4, 2)$, which results in the y-axis running from lower left to upper right.) Far away from the origin, the graph resembles a plane which is tilting upward in the direction of the positive y-axis. We can see these features better by cutting down the domain, increasing `PlotRange` and `PlotPoints`, moving the `ViewPoint`, and labeling the axes.

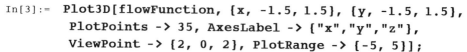

```
In[3]:= Plot3D[flowFunction, {x, -1.5, 1.5}, {y, -1.5, 1.5},
            PlotPoints -> 35, AxesLabel -> {"x","y","z"},
            ViewPoint -> {2, 0, 2}, PlotRange -> [-5, 5]];
```

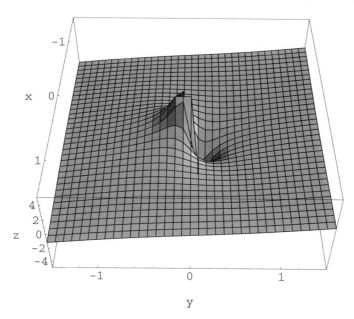

Graphing Level Curves

The second way to visualize a function of two variables is by means of a two-dimensional plot of its *level curves* or *contours*: that is, the curves along which the function takes a constant value. In *Mathematica*, such a plot is produced with the `ContourPlot` command. Here is an important example to keep in mind. If $g(x, y)$ denotes the elevation of a point (x, y) on a small region of the Earth, then the graph of g is the shape of the Earth's surface, and a plot of the level curves of g is a topographic map (with contour lines). You will often find topographic maps with different elevation ranges shaded in different colors. Similarly, *Mathematica* usually shades the regions between level curves to indicate different ranges of values of the function. Let's try our `flowFunction` example.

```
In[4]:= ContourPlot[flowFunction, {x, -2, 2}, {y, -2, 2}];
```

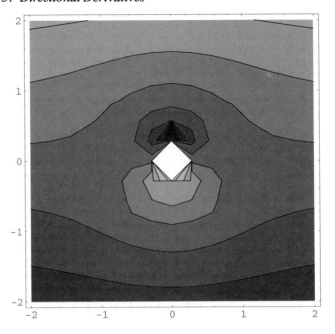

This is again not the greatest picture, but it does tell us quite a bit about the function f. The lightest regions of the plot correspond to the largest values of f, and the darkest to the smallest values. So, if we were to draw a contour plot of $-f$, the level curves would be the same, but the shading would be reversed, with the black and white regions interchanged. We can see from looking carefully at the shading pattern that for a fixed level c, the equation $f(x, y) = c$ describes a level curve with two components, both of which cross the y-axis. One component is roughly horizontal, except for a somewhat rounder portion in the middle, and extends infinitely to the right and to the left. The other component is a small loop located near the origin, above the origin if c is negative, and below the origin if c is positive. We can verify this observation by having *Mathematica* do a computation:

```
In[5]:=  Solve[(flowFunction /. {x->0}) == c, y]
```

$$\text{Out[5]}= \left\{ \left\{ y \rightarrow \frac{1}{2} \left(c - \sqrt{4 + c^2} \right) \right\}, \left\{ y \rightarrow \frac{1}{2} \left(c + \sqrt{4 + c^2} \right) \right\} \right\}$$

This shows that each contour crosses the y-axis at two distinct places, with y-coordinates differing by $\sqrt{4 + c^2} \geq 2$.

To better resolve the shapes of the contours, we can redo the contour plot taking a higher value for **PlotPoints**. (The picture is especially nice in color, which you can't see here, since we've printed this book in black and white. You can see the color on your computer screen if you re-evaluate the Chapter 5 Notebook on the accompanying disk.) The function is changing most rapidly where the contours are closest together.

```
In[6]:= conplot = ContourPlot[flowFunction, {x, -2, 2},
        {y, -2, 2}, ColorFunction -> Hue,
        PlotPoints -> 50, Contours -> 50];
```

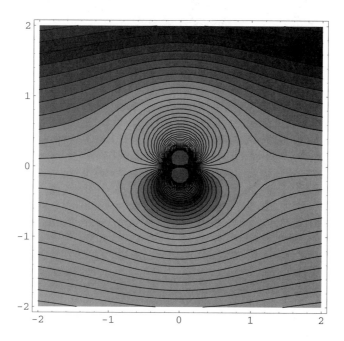

Although the "small loop" components of the contours appear to pass through the origin, they really don't, since the function f is undefined at this point.

Now we can explain the motivation for our example and the reason for the name **flowFunction**. If we restrict to the domain $x^2 + y^2 > 1$, then the level curves of our function are *streamlines*, that is, the trajectories along which objects would drift with the current, for an ideal fluid flowing around a circular obstacle in the shape of the disk $\{x^2 + y^2 \leq 1\}$.

This raises the issue of how to use **ContourPlot** to visualize a function whose domain does not fill up the whole rectangle enclosing it. Either we can redefine the function at the missing points, or else we can "block out" the portion of the region that is not supposed to be part of the domain. We do the latter for our example:

```
In[7]:= Show[conplot, Graphics[Disk[[0, 0], 1]]];
```

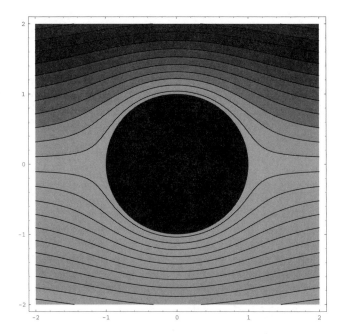

This is a nice picture of streamlines for flow around an obstacle (shown in black).

The Gradient of a Function of Two Variables

So far, we have discussed the level curves of a function, that is, the curves along which the function is not changing at all. Now we want to discuss how functions change from point to point. As in one-variable differential calculus, this will be measured by derivatives.

Partial Derivatives and the Gradient

There are several types of derivatives for a function f of two variables. It is natural to start with the easiest ones to compute: the *partial derivatives*. These are the derivatives of f viewed as a function of one variable while the other variable is held fixed. The partial derivatives of f with respect to x and y are denoted by f_x and f_y, or by $f^{(1,0)}$ and $f^{(0,1)}$, or by the Leibnitz notation $\partial f/\partial x$ and $\partial f/\partial y$. (Note the use of a curly ∂ in place of the ordinary d of one-variable calculus.) In *Mathematica*, the partial derivatives are computed using the D command. The second argument of D is the variable with respect to which one differentiates. For instance:

```
In[8]:=  D[x^3 + Sin[x*y], x]
```

$$\text{Out[8]= } 3x^2 + y\,Cos[x\,y]$$

Mathematica can also keep track of derivatives of functions that haven't been defined yet, as in:

```
In[9]:= D[unknownFunction[x, y], x]
```

$$\text{Out[9]}= \text{unknownFunction}^{(1,\,0)}[x, y]$$

It is particularly convenient to assemble the two partial derivatives into a single vector, called the *gradient* of f, and denoted with the special symbol ∇f. Thus

$$\nabla f = \left(\frac{\partial f}{\partial x}, \frac{\partial f}{\partial y}\right) = \frac{\partial f}{\partial x}\mathbf{i} + \frac{\partial f}{\partial y}\mathbf{j}. \tag{1}$$

As we will see shortly, *the gradient of a function at any point is perpendicular to the level curve through that point.* In a sense, this is not surprising, for a function is constant along level curves, and the gradient is concocted from derivatives, which measure change in a function. (Keep in mind the motto: "Change is orthogonal to constancy.")

Mathematica has two commands that are useful in the study of gradients, `Grad` (in the package `Calculus`VectorAnalysis`), and `PlotGradientField` (in the package `Graphics`PlotField`). To use the `Grad` command, first load the `VectorAnalysis` package and use `SetCoordinates` to tell *Mathematica* what your coordinates are called. (The usual choice will be `Cartesian[x, y, z]`.) Then `Grad` computes the gradient of a function as a three-dimensional vector, in other words, as a three-element list. To write the gradient of a function of two variables as a two-element list, apply `Drop` to remove the third entry (which will be 0, anyway). Here's an example:

```
In[10]:= <<Calculus`VectorAnalysis`
```

```
In[11]:= SetCoordinates[Cartesian[x, y, z]]
```

$$\text{Out[11]}= \text{Cartesian}[x, y, z]$$

```
In[12]:= Drop[Grad[flowFunction], -1]
```

$$\text{Out[12]}= \left\{\frac{2xy}{(x^2+y^2)^2}, \frac{1+2y^2}{(x^2+y^2)^2} - \frac{1}{(x^2+y^2)}\right\}$$

The `PlotGradientField` command produces a graphical depiction of the gradient, with arrows showing the direction and magnitude of the gradient at various points. Let's give an example using `flowFunction`.

```
In[13]:= <<Graphics`PlotField`
```

```
In[14]:= PlotGradientField[flowFunction, {x, -2, 2},
            {y, -2, 2}];
```

```
Power::infy : Infinite expression  1/0
  encountered. ∞::indet :
   Indeterminate expression 0 ComplexInfinity
  encountered.
```

```
Power::infy : Infinite expression  1/0²
encountered. ∞::indet :
 Indeterminate expression 0 ComplexInfinity
 encountered.
Power::infy : Infinite expression  1/0
encountered. General::stop : Further output
of Power::infy will be suppressed during
this calculation.
```

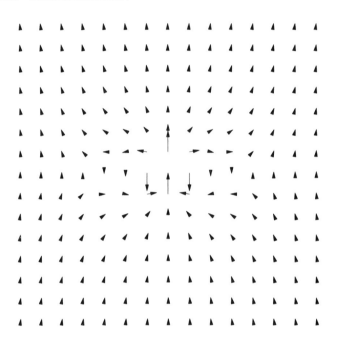

The error messages here are all due to the fact that the function and its gradient are not defined at the point $(0, 0)$. *Mathematica* is unhappy about this, but goes ahead anyway. In the future, we'll just suppress such messages, and you will eventually learn when to ignore them.

With the default settings, it often happens that many of the arrows are too short to be legible. You can improve their appearance by rescaling the arrows with the options `ScaleFactor` and `ScaleFunction`. It's best to use a `ScaleFunction` which is increasing, but which levels off somewhat for large values of the argument. Here is the same example after rescaling, displayed along with the contour plot to illustrate that the gradient vectors are perpendicular to the level curves. (For the use of the `DisplayFunction` option, see the Glossary to Problem Set B.)

```
In[15]:= gradplot = PlotGradientField[flowFunction,
         {x, -2, 2}, {y, -2, 2}, ScaleFunction -> (Tanh[#]&),
         DisplayFunction -> Identity];
```

```
In[16]:= conplot1 = Show[conplot, ContourShading -> False,
         DisplayFunction -> Identity];
```

```
In[17]:= Show[conplot1, gradplot, DisplayFunction ->
         $DisplayFunction];
```

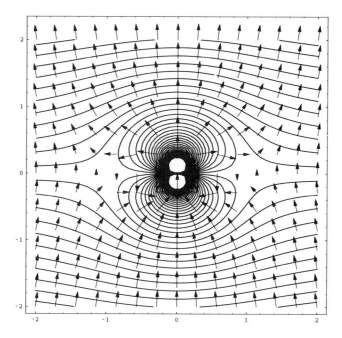

As you can see, the arrows representing the gradient are perpendicular to the level curves.

Directional Derivatives

The importance of the gradient comes from the formula for *directional derivatives*. Suppose f is a function of two variables and $t \mapsto \mathbf{r}(t) = (x(t), y(t))$ parametrizes a curve in the domain of f through a certain point, say $(x_0, y_0) = \mathbf{r}(0)$. Then we can form the composite function

$$g(t) = f \circ \mathbf{r}(t) = f(x(t), y(t)).$$

The chain rule for derivatives says that

$$g'(0) = \frac{\partial f}{\partial x}(x_0, y_0) \cdot x'(0) + \frac{\partial f}{\partial y}(x_0, y_0) \cdot y'(0) = \nabla f(x_0, y_0) \cdot \mathbf{r}'(0). \qquad (2)$$

Why are there two terms, and not just one? Well, a change in t affects the value of g in two different ways: once by changing $x(t)$ (at the rate $x'(t)$), which then affects f at the rate $\partial f/\partial x$; and once by changing $y(t)$ (at the rate $y'(t)$), which then affects f at the rate $\partial f/\partial y$. But note the vector form of the equation above: the rate of change of g is just the dot product of the gradient of f with the velocity vector of the curve. This suggests defining the *directional derivative* of f (at (x_0, y_0)) in the direction $\mathbf{u} = (u_0, u_1)$ to be

$$D_{\mathbf{u}}f(x_0, y_0) = \nabla f(x_0, y_0) \cdot \mathbf{u} = \frac{d}{dt}\bigg|_{t=0} f\left(x_0 + tu_0, y_0 + tu_1\right). \qquad (3)$$

(Although some books require $\mathbf{u}$ to be a unit vector here, there is no reason to make this restriction.) Then formula (2) becomes

$$\frac{d}{dt}\bigg|_{t=0} f(\mathbf{r}(t)) = D_{\mathbf{r}'(0)}f(\mathbf{r}(0)). \qquad (4)$$

Suppose the curve parametrized by $\mathbf{r}(t)$ is a level curve of f. By definition of the term "level curve", $f(\mathbf{r}(t))$ is a constant, so the left-hand side of equation (4) is 0. This tells us that $\nabla f(x_0, y_0) \cdot \mathbf{r}'(0) = 0$, i.e., the gradient of f at (x_0, y_0) is perpendicular to a tangent vector to the level curve through that point, and thus to the level curve itself. This explains the picture produced by input lines In[15] through In[17].

The directional derivative formula has other consequences. For instance, if we fix x_0 and y_0, and maximize $D_{\mathbf{u}}f(x_0, y_0)$ over all unit vectors $\mathbf{u}$, then the maximum value is $\|\nabla f(x_0, y_0)\|$, and is achieved when $\mathbf{u}$ and ∇f point in the same direction. This fact is often stated informally as follows: *the function f increases the fastest in the direction of the gradient vector.* We will use this fact in Chapter 7, *Optimization in Several Variables*.

We can see a final consequence of the directional derivative formula if we look at a *critical point* of f, that is, at a point (x_0, y_0) where $\nabla f(x_0, y_0) = \{0, 0\}$. At such a point, all directional derivatives of f are 0, i.e., the rate of change of f is 0 in all directions. The fact that the gradient is perpendicular to the level curve through such a point gives no information; indeed, the level curve through a critical point may not have a well-defined tangent direction at this point. We can see this in the example above. From Out[12], ∇f vanishes at $(\pm 1, 0)$, where the function f takes the value 0. Also, the level curve $f = 0$ is quite unusual in shape.

```
In[18]:= ContourPlot[flowFunction, {x, -2, 2},
            {y, -2, 2}, ContourShading -> False,
            PlotPoints -> 100, Contours -> {0}];
```

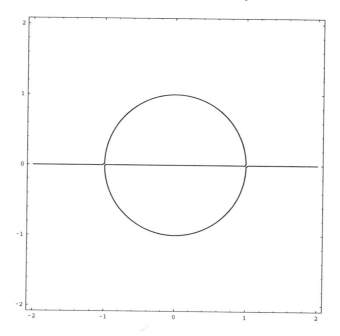

We see (as we can also check from the original formula for f) that the curve $f = 0$ is the union of the unit circle and the x-axis (with the origin deleted), and that the two pieces of the level curve cross each other at the critical points of f.

Functions of Three or More Variables

The theory of gradients and directional derivatives generalizes without difficulty to functions of three of more variables. Visualizing such functions, however, is more difficult. A function of three variables f is a rule that associates a value $f(\mathbf{x}) = f(x, y, z)$ to any $\mathbf{x} = (x, y, z)$ in a subset of $\mathbf{R}^3$ called the *domain* of the function. We will again mostly be interested in the case where f is *smooth*, i.e., has continuous partial derivatives. The set where such a function takes a fixed value c will now usually be a surface, called a *level surface* of the function. There is a *Mathematica* package `Graphics'ContourPlot3D'` that can be used to draw a picture of such a surface. Use it with caution, since with a large value of `PlotPoints`, the calculations can be incredibly slow. Here is a famous example of a level surface, the *hyperboloid of two sheets*.

```
In[19]:= <<Graphics'ContourPlot3D'
```

```
In[20]:= quadricFunction = z^2 - x^2 - y^2;
```

```
In[21]:= cp3d = ContourPlot3D[quadricFunction, {x, -1.1, 1.1},
         {y, -1.1, 1.1}, {z, -2, 2}, Contours -> {1.0},
         Axes -> True, AxesLabel -> ["x","y","z"]];
```

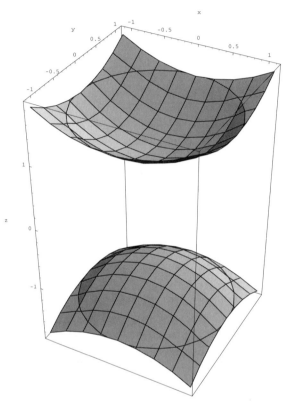

Other level surfaces of the same function $z^2 - x^2 - y^2$ include the *double cone* $z^2 = x^2 + y^2$ and the *hyperboloid of one sheet* $z^2 - x^2 - y^2 = c$ with $c < 0$. `ContourPlot3D` does not do a particularly good job with these. We will display them instead with `ParametricPlot3D`, in the next chapter on geometry of surfaces. But let's note the important fact that the double cone is not smooth; it has a "singular point" at the origin. We will explain shortly why it is that a level surface of a smooth function can fail to be smooth at certain points.

The gradient of a function f of three variables is defined, in analogy with (1), by

$$\nabla f = \left(\frac{\partial f}{\partial x}, \frac{\partial f}{\partial y}, \frac{\partial f}{\partial z} \right) = \frac{\partial f}{\partial x}\mathbf{i} + \frac{\partial f}{\partial y}\mathbf{j} + \frac{\partial f}{\partial z}\mathbf{k}.$$

The directional derivative of f in the direction $\mathbf{u} = (u_0, u_1, u_2)$ is defined by the analogue of equation (3):

$$D_{\mathbf{u}}f(x_0, y_0, z_0) = \nabla f(x_0, y_0, z_0) \cdot \mathbf{u} = \frac{d}{dt}\bigg|_{t=0} f(x_0 + tu_0, y_0 + tu_1, z_0 + tu_2).$$

If $t \mapsto \mathbf{r}(t)$ parametrizes a curve in $\mathbf{R}^3$, equation (4) above still applies. So, if the curve $\mathbf{r}(t)$ stays in a level surface $f(x, y, z) = c$, we conclude as before that for any point (x_0, y_0, z_0) in the surface, $\nabla f(x_0, y_0, z_0)$ is again perpendicular to the curve. This being true for *any* curve in the level surface, we see that $\nabla f(x_0, y_0, z_0)$ is perpendicular to the level surface through (x_0, y_0, z_0). We can confirm this visually using the *Mathematica* command `PlotGradientField3D`, part of the `Graphics`PlotField3D`` package, to visualize the gradient vectors.

```
In[22]:= <<Graphics`PlotField3D`
```

```
In[23]:= vecplot3d = PlotGradientField3D[quadricFunction,
         {x, -1.1, 1.1}, {y, -1.1, 1.1},
         {z, -2, 2}, PlotPoints->3, VectorHeads -> True];
```

```
In[24]:= Show[vecplot3d, cp3d];
```

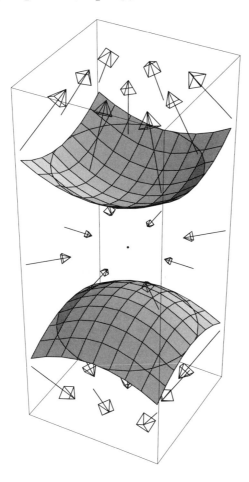

Now we can see why it is that the level surface $z^2 - x^2 - y^2 = 0$ (the double cone) has a singular point at the origin. The gradient of $z^2 - x^2 - y^2$ is $(-2x, -2y, 2z)$, which degenerates to $(0, 0, 0)$ at the origin. Thus we cannot use the gradient to find a unit vector perpendicular to the cone at the origin, and in fact the cone has *no* tangent plane at this point. On the other hand, if f is a smooth function and if $\nabla f(x_0, y_0, z_0) \neq (0, 0, 0)$, then the level surface $f(x, y, z) = f(x_0, y_0, z_0)$ has a well-defined tangent plane at (x_0, y_0, z_0), namely the unique plane through this point which is normal to the vector $\nabla f(x_0, y_0, z_0)$. So a level surface of a smooth function of three variables is *nonsingular* provided it doesn't pass through any *critical point* of the function (a point where the gradient vector vanishes). This is a higher-dimensional analogue of the fact that a level *curve* of a smooth function of *two* variables is smooth (without "crossing points") if it does not pass through a critical point.

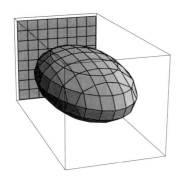

Problem Set E

DIRECTIONAL DERIVATIVES AND THE GRADIENT

1. This problem examines in detail the level curves of the function
$$f(x,\, y) = (x^2 + y^2)^2 - x^2 + y^2.$$

   (a) Graph f in the region $-2 \le x \le 2, -2 \le y \le 2$. Draw a contour plot in the same domain.

   (b) To focus on the level curves in greater detail, turn off `ContourShading`, and draw three plots, each with a different `ContourStyle`, one with `Contours` set to `{-0.2, -0.1}`, one with `Contours->{0}`, and one with `Contours` set to `{0.1, 0.2, 1, 4, 10}`. To get a good picture, especially for the contour $f = 0$, you may need to set `PlotPoints` as high as 150. Then put all the contour plots back together with `Show`. You should see three distinct shapes of level curves: one shape for negative values of f, one shape for positive values of f, and one special shape for $f = 0$. Can you explain what is responsible for the differences? (If not, just state what the three shapes are and come back to this question after you've done the next parts of the problem.)

   (c) Compute the gradient of f, and determine where it vanishes. (Hint: Apply `Solve` to the equation setting the gradient equal to the zero-vector. You need to use `{0,0}` if you think of the gradient as a two-element list, and `{0,0,0}` if you think of the gradient as a three-element list.) What is special about the level curves through the critical points?

(d) Use `PlotGradientField` to draw the gradient field of f, and superimpose the gradient plot on your contour plot. What do you observe about the gradient vectors along the three different kinds of level curves? What happens at the origin, and how does this help explain the special shape of the level curve $f = 0$? Complete your answer to (b).

2. Some pairs of functions of two variables have the property that the level curves of one function are orthogonal to the level curves of the other. This problem investigates that phenomenon.

(a) Suppose that u and v are smooth functions of x and y and that

$$\frac{\partial u}{\partial x} = \frac{\partial v}{\partial y}, \qquad \frac{\partial u}{\partial y} = -\frac{\partial v}{\partial x}.$$

Verify that $\nabla u \cdot \nabla v = 0$ and that u and v are both *harmonic functions*; that is, they each satisfy *Laplace's equation*:

$$\frac{\partial^2 u}{\partial x^2} + \frac{\partial^2 u}{\partial y^2} = \frac{\partial^2 v}{\partial x^2} + \frac{\partial^2 v}{\partial y^2} = 0.$$

(You will need the fact, found in all multivariable calculus texts, that for a smooth function, say u, $\dfrac{\partial^2 u}{\partial x \partial y} = \dfrac{\partial^2 u}{\partial y \partial x}$.) Under these circumstances, u and v are said to be *harmonic conjugates* of one another.

(b) Verify that the following pairs of functions are harmonic conjugates of one another. In each case, plot the level curves, using a different `ContourStyle` for each function, and then superimpose the two contour plots with `Show` to give a picture of the orthogonal families of level curves. (You may need to experiment with the domain of the plot until you get a good picture. You may also need to increase the settings for `Contours` and `PlotPoints`. Use a different `ContourStyle` for each family of curves to distinguish the two families.)

(i) $u(x, y) = x$, $v(x, y) = y$ (the simplest case).

(ii) $u(x, y) = x^2 - y^2$, $v(x, y) = 2xy$ (hyperbolas).

(iii) $u(x, y) = \sin x \cosh y$, $v(x, y) = \cos x \sinh y$.

(iv) $u(x, y) = x\left(1 + \dfrac{1}{x^2 + y^2}\right)$, $v(x, y) = y\left(1 - \dfrac{1}{x^2 + y^2}\right)$.

(v) $u(x, y) = \dfrac{1}{2}\ln\left(x^2 + y^2\right)$, $v(x, y) = \arctan\left(\dfrac{y}{x}\right)$, $x > 0$.

3. Consider a conical mountain, rising from a flat plain, in the shape of the graph of the height function $h(x, y) = \max(3(1 - \sqrt{x^2 + y^2}), 0)$. (The "max" is put in

to cut the function off at 0, since away from the mountain the height should be everywhere 0.)

(a) Draw a picture of the mountain using `Plot3D`, and verify that it has an appropriate "mountain-like" appearance. Save the result as a `Graphics3D` object called `mountain`, since you will need it later in the problem.

(b) Now suppose that a builder wants to build a road up the mountain to the top. The road may be idealized as a curve lying on the mountain, so it has parametrization

$$(\mathbf{r}(t),\, z(t)), \qquad z(t) = h(\mathbf{r}(t)),$$

where $\mathbf{r}(t) = (x(t),\, y(t))$ is the parametric equation of the projection of the road into the x-y plane. It's not practical for the road to go straight up the side of the mountain, since it's too steep. Therefore, the functions $x(t)$ and $y(t)$ should be chosen so that the directional derivative of h in the direction of a unit vector pointing along the road, which represents the rate of climb of the road, is a fixed reasonable constant, say $\frac{1}{20}$. (This constant corresponds to a 5% grade, quite a steep road by normal standards.) This condition amounts to a differential equation for $x(t)$ and $y(t)$. We may as well take the curve $\mathbf{r}(t)$ to be parametrized by arclength, so that $x'(t)^2 + y'(t)^2 = 1$. Show then that the 5% grade condition is satisfied by the spiral

$$x(t) = at\cos(b\ln t), \quad y(t) = at\sin(b\ln t), \quad z(t) = 3(1 - at), \text{ for } 0 \le t \le \frac{1}{a},$$

when $a = \frac{1}{60}$, $b = \sqrt{3599}$.

(c) Use `ParametricPlot3D` to plot the spiral, getting a `Graphics3D` object called `road`. Then superimpose it on the picture of the mountain to draw the road going up the mountain. This takes some fiddling since you need the plot of the road to be thick enough to show up against the backdrop of the mountain. You can achieve this is with the command

```
Show[mountain, Graphics3D[AbsoluteThickness[2]], road]
```

in which the `AbsoluteThickness` directive orders the curve to be printed thicker when `road` is displayed. (Hint: With three-dimensional graphics, when one object lies in front of another, *Mathematica* only displays the one in front. This creates a problem when we try to display simultaneously a surface and a curve lying on that surface, because round-off errors in the calculations may cause the surface to appear to lie in front of the curve a good bit of the time, thereby obscuring the curve. You can fix this either by sliding the curve forward or the surface back.)

4. In this problem, we further explore the notions of directional derivative and of derivative along a curve. Consider the function of two variables

$$f(x, y) = x^3 - 2xy^2 + xy - y^2.$$

(a) Begin by applying `ContourPlot` and `Plot3D` to the function in the square domain $\{-3 \leq x \leq 3, -3 \leq y \leq 3\}$. (Increase `PlotPoints` and/or `Contours` until you have reasonable pictures.) Compute the gradient of f and locate the critical points. (See Problem 1 for a method to do this.) After throwing away the complex critical points, plot the critical points with `ListPlot` (it helps to set the `PlotStyle` to $\{$`PointSize[0.03]`$\}$ so that the dots are big enough to be easily visible) and superimpose them on your contour plot.

(b) Next, let's restrict f to lines in the x-y plane, and investigate the curves in the graph of f lying over these lines. Interesting cases are the lines $2x - 3y = 0$ and $3x - 5y + 9 = 0$. Superimpose these lines on your contour plot so you can see roughly how f behaves along them. Also use `ParametricPlot3D` to draw the curves in the graph of f lying over these lines, and superimpose these on your surface plot. (See part (c) in the previous problem for a method to do this effectively.)

(c) Consider the functions $g(x) = f(x, 2x/3)$ and $h(x) = f(x, (3x + 9)/5)$. These are the restrictions of f to the lines in part (b). Plot g and h and analyze their behavior. Then check that

$$g'(x) = (D_{(1,2/3)}f)(x, 2x/3), \qquad h'(x) = (D_{(1,3/5)}f)(x, (3x + 9)/5),$$

and explain why this should be the case from the directional derivative formula (formula (4) in Chapter 5).

(d) Finally, let's study what happens when we lift one of the level curves of f to the graph of f. By definition, f is constant along a level curve, so the lift of the level curve to the graph should be a curve of constant height. One way to visualize such a curve is to intersect the surface with a horizontal plane, by displaying the three-dimensional graph of f and the plane simultaneously. Do this for the horizontal planes $z = 0$ and $z = 10$, and check that the shape of the intersections agrees with what you get from contour plots with `Contours` set to $\{0\}$ and $\{10\}$, respectively. The level curve $f = 0$ is special; explain how and why.

5. In this problem, we study the tangent plane to a surface at a point. We know that for a function f of three variables, ∇f points perpendicular to the level surfaces of f. Thus, if (x_0, y_0, z_0) is not a critical point of f, then the level surface

$$f(x, y, z) = f(x_0, y_0, z_0)$$

has a well-defined tangent plane at (x_0, y_0, z_0); namely, the plane through this point with $\nabla f(x_0, y_0, z_0)$ as a normal vector. For each of the following functions f, draw a picture of the level surface through the indicated point (using `ContourPlot3D`), compute the equation of the tangent plane at this point, and display the tangent plane and the level surface simultaneously to visualize the tangency. (These are examples of *quadric surfaces*, i.e., surfaces given by quadratic equations.)

(a) elliptic paraboloid: $f(x, y, z) = x^2 + 4y^2 - z$, $(x_0, y_0, z_0) = (0, 0, 0)$.

(b) hyperbolic paraboloid or saddle surface: $f(x, y, z) = x^2 - 4y^2 - z$, $(x_0, y_0, z_0) = (0, 0, 0)$.

(c) ellipsoid: $f(x, y, z) = x^2 + 4y^2 + 9z^2$, $(x_0, y_0, z_0) = (1, 0, 0)$.

(d) $f(x, y, z) = xy + xz$, $(x_0, y_0, z_0) = (1, 1, 0)$. See if you can identify the surface.

6. For each of the following functions f of three variables, plot the gradient field using `PlotGradientField3D`, and use `ContourPlot3D` to visualize the indicated level surfaces. (You may need to experiment with the `ViewPoint` and `PlotPoints`, and with the ranges of x, y, and z values to get nice pictures.) Superimpose the two plots to see that each level surface is perpendicular to the gradient field.

(a) $f(x, y, z) = x^2 + y^2 + z^2$, $f = 1$ (a sphere).

(b) $f(x, y, z) = x^3 - 3xy^2 + z$, $f = 0$ (the "monkey saddle").

(c) $f(x, y, z) = xyz$, $f = 1$.

7. Given a function f of two or three variables, it is interesting to study curves tangent to the gradient vector field of f. Let's make this precise. For simplicity let's restrict to the two-variable case. A curve C with parametrization $t \mapsto \mathbf{r}(t) = (x(t), y(t))$ is everywhere tangent to the gradient vector field of f if for all values of t, $\mathbf{r}'(t)$ is a scalar multiple of $\nabla f(\mathbf{r}(t))$. Since we are free to reparametrize, we may as well take the scalar multiple to be 1, so we obtain the vector-valued differential equation

$$\mathbf{r}'(t) = \nabla f(\mathbf{r}(t)),$$

or the pair of coupled scalar-valued differential equations

$$\begin{cases} x'(t) = f_x(x(t), y(t)), \\ y'(t) = f_y(x(t), y(t)). \end{cases}$$

These equations are then said to give the *gradient flow* of the function f, and C is called an *integral curve* of the flow. At each noncritical point $(x(t), y(t))$ on such a curve C, the unit tangent vector to C is parallel to $\nabla f(x(t), y(t))$, hence orthogonal to the level curve through this point. So the level curves and the curves obtained from the gradient flow form two mutually orthogonal families. (See Problem 2.)

For each function below, draw a contour plot of the function. Then solve the differential equations for the gradient flow through the indicated points, by using either DSolve or NDSolve, as appropriate. Plot the integral curves of the gradient flow in a different style from the level curves of the contour plot, and then superimpose them to view the two orthogonal families.

(a) $f(x, y) = x^2 - y^2$ in the region $-2 \leq x \leq 2$, $-2 \leq y \leq 2$. Plot the integral curves through the points

$$\left(\cos\left(\frac{j\pi}{12} \right), \ \sin\left(\frac{j\pi}{12} \right) \right), \qquad 0 \leq j \leq 24,$$

which are equally spaced around the unit circle. (In this case DSolve can easily solve the differential equation with arbitrary initial conditions, so do this first and then substitute.)

(b) $f(x, y) = \cos x + \sin y$ in the region $-\pi \leq x \leq \pi$, $-\pi \leq y \leq \pi$. Use DSolve to solve the equations for the integral curves starting at a point of the form $(a, 0)$. You will see more than one solution, but the solutions only differ in some signs, and it's not so hard to figure out what the appropriate solution is, going *forward* in time, for $-\pi < a < \pi$. Plot the integral curves through the points

$$\left(\frac{j\pi}{12}, 0 \right), \qquad -11 \leq j \leq 11,$$

for $0 \leq t \leq 4$, and superimpose them on the part of the contour plot with $0 \leq y \leq \pi/2$. After adjusting the AspectRatio, you should clearly see the orthogonal families.

(c) $f(x, y) = x^3 + 3xy + y^3$ in the region $-2 \leq x \leq 2$, $-2 \leq y \leq 2$. Locate the critical points; there are two of them. This time the differential equations are not very well behaved and you need to use NDSolve. The calculations may be very slow, especially when trajectories get close to the critical points, so plot only a few trajectories, some starting on the line $x = -2$, and some starting on $y = -2$. In both cases you want to go forward in time, say for $0 \leq t \leq 2$, to get integral curves lying in the desired region.

Glossary of Some Useful *Mathematica* Objects

Commands

`ContourPlot` Produces a contour plot of a function $f(x, y)$.

`ContourPlot3D` Plots a level surface $f(x, y, z) = c$.

`D` Computes the derivative or partial derivative.

`Drop` Drops one or more elements from a list.

`DSolve` Symbolic differential equation solver.

`Flatten` Removes extra parentheses from nested lists.

`Grad` Computes the gradient of a function.

`Graphics3D` For creating three-dimensional graphics objects.

`ListPlot` Plots individual points in 2-space.

`NDSolve` Numerical differential equation solver.

`ParametricPlot` Plots a curve parametrically in 2-space.

`ParametricPlot3D` Plots a curve parametrically in 3-space.

`Plot3D` Plots a surface $z = f(x, y)$.

`PlotGradientField` Plots the gradient field of a function of two variables.

`PlotGradientField3D` Plots the gradient field of a function of three variables.

`SetCoordinates` Sets the coordinate system for `Grad`, etc.

`Show` Displays several graphics objects simultaneously.

`Table` Creates a list.

Options and Directives

`AbsolutePointSize` Specifies point size in printer's points.

`AbsoluteThickness` Specifies line thickness in printer's points.

`ColorFunction` Specifies coloring scheme.

`ContourShading` Specifies whether to shade regions between contours.

`ContourStyle` Specifies style for `ContourPlot`.

`Contours` Specifies which (or how many) contours to show.

`PlotPoints` Specifies how many points to sample in a plotting routine.

`PlotStyle` Specifies style for `Plot` and `ListPlot`.

`PointSize` Specifies (relative) point size.

`RGBColor` Specifies a color. `RGBColor[1,0,0]` means red; `RGBColor[0,0, 1]` means blue, `RGBColor[0,1,0]` means green.

`ScaleFunction` Rescales arrows in the commands `PlotGradientField` and `PlotGradientField3D`.

`Thickness` Specifies (relative) line thickness.

`VectorHeads` Option in `PlotField3D`, specifying whether or not vectors should be drawn with arrowheads. If so, `PlotPoints` should be kept small.

`ViewPoint` Point from which to view three-dimensional graphics.

Packages

`Calculus`VectorAnalysis`` needed for `Grad`.

`Graphics`ContourPlot3D`` needed for `ContourPlot3D`.

`Graphics`PlotField`` needed for `PlotGradientField`.

`Graphics`PlotField3D`` needed for `PlotGradientField3D`.

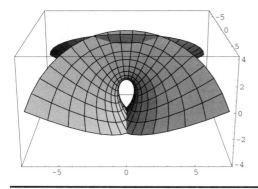

Chapter 6

GEOMETRY
OF SURFACES

Just as a curve is (at least locally) the image of a continuous function from an interval in $\mathbf{R}$ to $\mathbf{R}^3$, a *surface* is (at least locally) the image of a continuous function from a domain in $\mathbf{R}^2$ to $\mathbf{R}^3$. We shall carry out a program for surfaces in space similar to our study of curves in Chapter 3. Not surprisingly, the details are more complicated. Therefore we will be content on occasion to state results without elaborate justification. Our overall goals for surfaces are the same as they were for curves:

- to associate geometric invariants to surfaces;
- to describe and compute these invariants analytically using calculus; and
- to show that the invariants encode enough information about the geometric object to characterize it.

The main invariants we will investigate are the normal vector, the tangent plane, and various forms of curvature. You may think of the normal vector and tangent plane as analogs of the Frenet frame for a curve, and of the curvatures of a surface as the analogs of the curvature and torsion for a curve. Since surfaces are two-dimensional and curves only one-dimensional, there is a great deal more information to encode—thus accounting for increased complexity of the analytic formulas.

The Concept of a Surface

We shall define a surface by analogy with our definition of a curve. It is common to give a definition of surfaces that allows us to glue together several *patches*. To keep matters simple, however, we only deal with single patches. So, a surface Σ is the image of a continuous function $\sigma : D \to \mathbf{R}^3$, where D is either a rectangle or a disk in $\mathbf{R}^2$. The function σ is called a *parametrization* of the surface. If u and v are coordinates on D, then via σ we may think of them as parameters or *generalized coordinates* on Σ. As with curves, we shall be ambiguous about whether the boundary of D is included or not.

In order to proceed, we need to impose additional conditions on the function σ. First, we say that σ is *regular* at a point if its partial derivatives, σ_u and σ_v, are continuous, nonzero, and noncollinear there. (Points where σ is not regular are called *singular* points.) At a regular point, the level curves of u (with parametrization $v \mapsto \sigma(u_0, v)$) and of v (with parametrization $u \mapsto \sigma(u, v_0)$) provide a nice, cross-hatched family on the surface. Since the tangent vectors σ_v and σ_u to these curves do not lie on the same line, they must span a well-determined *tangent plane* at any regular point. The vector $\mathbf{N} = \sigma_u \times \sigma_v$ is normal to this tangent plane, and hence to the surface itself.

Next, we say that σ is a *patch* if it is regular and one-to-one with a continuous inverse function, except perhaps at some exceptional points. When there are no exceptional points, we say that σ is a *smooth patch*. On a smooth patch, the normal vector $\mathbf{N}$ is a nonzero continuous function, and thus the surface has an *orientation* in the sense that there is a preferred normal direction at each point. (This need not be true for surfaces composed of more than one patch.)

If we were to require our surface to be the image of a single smooth patch, then we would have to exclude certain standard (and important) surfaces such as the sphere and the cone. So, we allow our parametrizations to fail to be one-to-one or regular on a one-dimensional subset, just as we allow parametrizations for curves, such as the circle or the cycloid, to fail in the same way on a zero-dimensional subset (i.e., a discrete set of points). We defer a rigorous treatment of these matters to a more advanced course in Advanced Calculus, Differential Geometry, Topology, or Manifolds.

You have already encountered one important example of a patch. Consider a surface Σ that is the graph of a function φ of two variables. A *Monge patch* on Σ is given by the parametrization $\sigma(u, v) = (u, v, \varphi(u, v))$. In this case, σ is one-to-one with continuous inverse provided φ is continuous. Moreover,

$$\sigma_u = (1, 0, \varphi_u) \quad \text{and} \quad \sigma_v = (0, 1, \varphi_v)$$

are never collinear. So, regularity reduces to the assumption that the partial derivatives φ_u, φ_v are continuous.

Basic Examples

In this section, we discuss some basic examples of surfaces, and explain how to parametrize and plot them in *Mathematica*.

THE SPHERE. The unit sphere centered at the origin is parametrized by spherical coordinates:

$$\sigma(u, v) = (\cos u \sin v, \sin u \sin v, \cos v), \quad u \in [0, 2\pi], \quad v \in [0, \pi].$$

Here u is called the *longitude*, v the *colatitude*. The coordinate functions,

$$x = \cos u \sin v, \qquad y = \sin u \sin v, \qquad \text{and} \quad z = \cos v,$$

satisfy the equation $x^2 + y^2 + z^2 = 1$. Thus, the sphere is a *quadric surface*— a surface determined by a quadratic equation in x, y, and z. (Other quadric surfaces were plotted in Problem Set E, Problem 5.) We can plot the sphere using `ParametricPlot3D`.

```
In[1]:= sphere = {Cos[u] Sin[v], Sin[u] Sin[v], Cos[v]};
```

```
In[2]:= ParametricPlot3D[Evaluate[sphere],
          {u, 0, 2Pi}, {v, 0, Pi}, ViewPoint -> {1, 1, 1},
          Axes -> False, Boxed -> False];
```

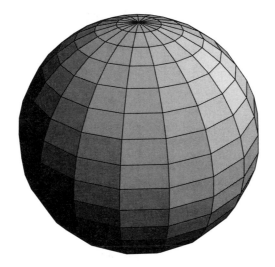

A SINUSOIDAL CYLINDER. Any surface parametrized in the form

$$\sigma(u, v) = \mathbf{r}(u) + v\mathbf{q},$$

where $\mathbf{r}(u)$ is a plane curve and $\mathbf{q}$ is a fixed direction, perpendicular to the plane of $\mathbf{r}$, is called a *cylindrical surface*. The sinusoidal cylinder is a cylindrical surface determined by the usual sine curve in the x-y plane and the perpendicular z-direction:

$$\sigma(u, v) = (u, \sin u, v), \quad u \in [-2\pi, 2\pi], \quad v \in [-1, 1].$$

We again plot the surface using **ParametricPlot3D**.

```
In[3]:=  sinusoid = {u, Sin[u], v};
```

```
In[4]:=  ParametricPlot3D[Evaluate[sinusoid],
          {u, -2Pi, 2Pi}, {v, -1, 1}, ViewPoint -> {1, 1, 1},
          Axes -> False, Boxed -> False];
```

A CONE. The cone is an example of a *ruled surface*; that is, one swept out by a straight line moving along a base curve. The curve need not be planar and the direction of the line may change. Still, a ruled surface must have a parametrization of the form

$$\sigma(u, v) = \mathbf{r}(u) + v\mathbf{q}(u),$$

where $\mathbf{r}$ is the curve along which the line is traveling and $\mathbf{q}$ is the direction of the line. (The line through the point $\mathbf{r}(u)$ is parametrized by $v \mapsto \mathbf{r}(u) + v\mathbf{q}(u)$.) Clearly, a cylindrical surface is ruled. The right circular cone may be parametrized as follows:

$$\sigma(u, v) = ((1 + v)\cos u, (1 + v)\sin u, 1 + v), \quad u \in [0, 2\pi], \quad v \in \mathbf{R}.$$

You should recognize this surface as a portion of the quadric surface $x^2 + y^2 - z^2 = 0$. The base curve is $\mathbf{r}(u) = (\cos u, \sin u, 1)$, and the line at the point $\mathbf{r}(u)$ is the one through the origin given by $v \mapsto (v + 1)\mathbf{r}(u)$. We plot the double cone using this parametrization.

```
In[5]:=  cone = {(1+v)*Cos[u], (1+v)*Sin[u], 1+v};
```

```
In[6]:= ParametricPlot3D[Evaluate[cone],
        {u, 0, 2Pi}, {v, -2, 0}, ViewPoint -> {1, 1, 1},
        Axes -> False, Boxed -> False];
```

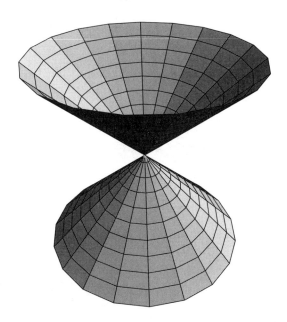

(The `Graphics`Shapes`` package in *Mathematica* contains, among other useful objects, a `Cone` command that can be used to generate a (single) cone, with its ruling, as a `Graphics3D` object.)

THE MONKEY SADDLE. The graph of the function $z = x^3 - 3xy^2$ provides an example of a Monge patch:

$$\boldsymbol{\sigma}(u, v) = (u, v, u^3 - 3uv^2), \quad u \in [-2, 2], \quad v \in [-2, 2].$$

The easiest way to visualize a Monge patch is with `Plot3D`:

```
In[7]:= Plot3D[u^3 - 3*v^2, {u, -2, 2}, {v, -2, 2},
        ViewPoint -> {1, 2, 0.5}, Axes -> False,
        Boxed -> False, AspectRatio -> 1.5];
```

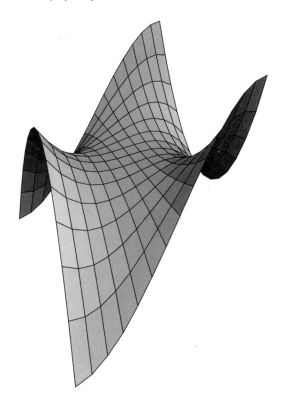

THE TORUS. The torus is an example of a *surface of revolution*; that is, a surface obtained by revolving a plane curve about an axis that does not meet the curve. To simplify, assume the plane curve lies in the y-z plane, and that it is revolved around the z-axis. Then, if $(0, g(u), h(u))$ parametrizes the curve, a parametrization of the surface can be given by

$$\boldsymbol{\sigma}(u, v) = (g(u)\cos v, \, g(u)\sin v, \, h(u)).$$

If we start with the circle $(y - 2)^2 + z^2 = 1$, then

$$\boldsymbol{\sigma}(u, v) = ((2 + \cos u)\cos v, (2 + \cos u)\sin v, \sin u), \quad u \in [0, 2\pi], \quad v \in [0, 2\pi],$$

and the resulting surface of revolution is a torus. *Mathematica* gives the picture:

```
In[8]:= torus={(2+Cos[u])Cos[v],(2+Cos[u])Sin[v],Sin[u]};

In[9]:= ParametricPlot3D[Evaluate[torus],
        {u, 0, 2Pi}, {v, 0, 2Pi}, ViewPoint -> {4, 4, 2},
        Axes -> False, Boxed -> False];
```

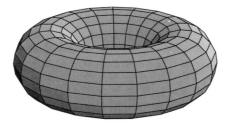

THE HELICOID. The helicoid, obtained by connecting the points on the right circular helix horizontally to the z-axis, is another example of a ruled surface.

$$\boldsymbol{\sigma}(u, v) = (v \cos u, v \sin u, u), \quad u \in [0, 3\pi], \quad v \in [0, 3].$$

```
In[10]:= helicoid = {v*Cos[u], v*Sin[u], u};
```

```
In[11]:= ParametricPlot3D[Evaluate[helicoid],
        {u, 0, 3Pi}, {v, 0, 3}, ViewPoint -> {4, 0, 2},
        Axes -> False, Boxed -> False];
```

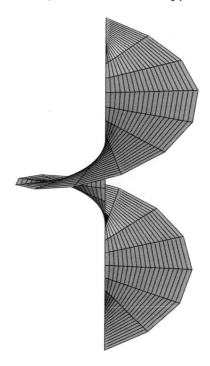

THE HYPERBOLOID OF ONE SHEET. In the last chapter, we looked at the hyperboloid of two sheets, given implicitly by the equation $x^2 + y^2 - z^2 = -1$. We deferred to this chapter the hyperboloid of one sheet, given by $x^2 + y^2 - z^2 = 1$. The one-sheeted hyperboloid is both a quadric surface and a ruled surface. A ruled parametrization

$$\sigma(u, v) = \mathbf{r}(u) + v\mathbf{q}(u),$$

can be given by taking

$$\mathbf{r}(u) = (\cos u, \sin u, 0), \qquad \mathbf{q}(u) = (-\sin u, \cos u, 1),$$

or else by pairing the same function $\mathbf{r}(u)$ with $\mathbf{q}(u) = (\sin u, -\cos u, 1)$. (The two rulings are mirror images of one another.) We check using *Mathematica* that the parametrization satisfies the equation of the hyperboloid, and then sketch the surface.

```
In[12]:= hyperboloid = {Cos[u], Sin[u], 0} +
         v{-Sin[u], Cos[u], 1};
```

```
In[13]:= hyperboloid[[1]]^2 + hyperboloid[[2]]^2 -
         hyperboloid[[3]]^2 // Simplify
```

```
Out[13]= 1
```

```
In[14]:= ParametricPlot3D[Evaluate[hyperboloid],
         {u, 0, 2Pi}, {v, -2, 2}, Axes -> False,
         Boxed -> False];
```

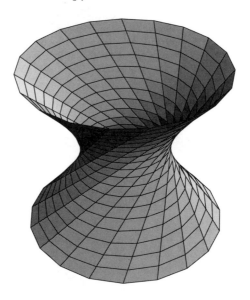

The Implicit Function Theorem

As we saw in Chapter 5, *Directional Derivatives*, surfaces can also be described as the locus of solutions to an equation $f(x, y, z) = c$. In order to understand how, at least locally, the solution set is a surface in the sense of this chapter, we must understand the Implicit Function Theorem. Let us first consider it for curves and then explain the generalization to surfaces.

Suppose we have a function $f(x, y)$ and we consider a level curve $f(x, y) = c$. Typically, such curves look like

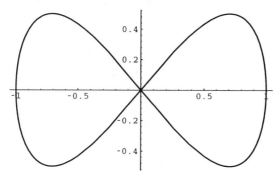

We would like to know which points (x_0, y_0) on the curve have the property that we can find exactly one smooth function $y = \varphi(x)$, defined in some interval I containing x_0 in its interior, for which $f(x, \varphi(x)) = c$, for all $x \in I$. Clearly, for the curve depicted, this will not be the case at the vertical and the crossing points, but it will be true elsewhere. Here is a basic theorem that answers the question.

Implicit Function Theorem. *Suppose that f, f_x, and f_y are continuous in a neighborhood of (x_0, y_0). Moreover, suppose that $f(x_0, y_0) = c$ and $f_y(x_0, y_0) \neq 0$. Then there is an open interval I containing x_0 such that for every $x \in I$ there is a unique number y for which $f(x, y) = c$. The function $y = \varphi(x)$ thus defined, and its derivative, are continuous functions on I.*

Idea of Proof. If $f_y(x_0, y_0) \neq 0$, then it is either positive or negative. Without loss of generality we may assume it is positive (otherwise replace f by $-f$). Since f_y is continuous, we may assume f_y is positive on some disk containing (x_0, y_0). In particular, if $F(x, y) = f(x, y) - c$, then $y \mapsto F(x_0, y)$ is an increasing function that vanishes at (x_0, y_0). So, there is a small positive number k for which

$$F(x_0, y_0 + k) > 0 \quad \text{and} \quad F(x_0, y_0 - k) < 0.$$

By continuity of F, we can find another small number h so that

$$F(x, y_0 + k) > 0 \quad \text{and} \quad F(x, y_0 - k) < 0 \quad \text{for} \quad |x - x_0| < h.$$

For any x in this small interval, by the Intermediate Value Theorem, there must be a value of y for which $F(x, y) = 0$. Since this value depends on x, we denote it by $y = \varphi(x)$. Moreover, because the function F_y is positive on the original disk, the value $\varphi(x)$ is uniquely determined by x. Thus we have found the desired function.

We omit the proof of continuity for φ and its derivative. (See any Advanced Calculus textbook.)

Now it is easy to see how the result generalizes. We state it without proof.

Implicit Function Theorem (for Three Variables). *Suppose that $f(x, y, z)$ and f_x, f_y, f_z are continuous on a neighborhood of (x_0, y_0, z_0). If $f(x_0, y_0, z_0) = c$ and $f_z(x_0, y_0, z_0) \neq 0$, then there is an open disk D containing (x_0, y_0) such that for every $(x, y) \in D$ there is a unique number $z = \varphi(x, y)$ for which $f(x, y, \varphi(x, y)) = c$. The parametrization $\sigma(u, v) = (u, v, \varphi(u, v))$, $(u, v) \in D$, thus defined, yields a smooth Monge patch surface on which the function f takes only the value c.*

Similarly, we can show that if $f(x, y, z)$ is a continuously differentiable function, if $f(x_0, y_0, z_0) = c$, and if $\nabla f(x_0, y_0, z_0) \neq 0$, then the level surface $f(x, y, z) = c$ is indeed a surface in the sense of this chapter in the vicinity of (x_0, y_0, z_0), even though it may not have a Monge patch parametrization.

Geometric Invariants

Now we begin to develop geometric invariants associated with surfaces. Let $\sigma(u, v)$ be defined on a domain D and parametrize a smooth surface Σ. The vectors σ_u, σ_v are, by hypothesis, noncollinear. Each is tangent to a curve on the surface, so they span a plane, called the *tangent plane*. Now any curve on the surface corresponds to a curve in D. Here is where we use the inverse continuity property of the parametrization! Thus, curves on the surface may be parametrized by $t \mapsto \sigma(\mathbf{r}(t))$, where $\mathbf{r}(t) = (u(t), v(t))$ is a plane curve inside D. Then a tangent vector to the surface is obtained by differentiating with respect to t. By the chain rule, we obtain

$$\frac{d}{dt}\sigma(\mathbf{r}(t)) = \sigma_u u'(t) + \sigma_v v'(t).$$

That is, any tangent vector is a linear combination of σ_u and σ_v, and our notion of the tangent plane is a good one. We shall always write

$$\mathbf{U}(u, v) = \frac{\sigma_u(u, v) \times \sigma_v(u, v))}{\|\sigma_u(u, v) \times \sigma_v(u, v)\|};$$

it is a unit normal vector to the surface at the point $\sigma(u, v)$. It is a *first-order* object since it involves first derivatives. In order to characterize curves, we needed second-order (curvature) and third-order (torsion) objects. Because of the extra degrees of freedom in space, we only need second-order objects to characterize surfaces. But we shall see that there are several different kinds of curvature that we can associate to a surface. More than one of them will be required for characterization.

All of the curvatures arise from differentiating the normal vector. The normal vector is an example of a *vector field*, a vector-valued function defined at each point of the surface. We can differentiate a vector field in any direction on the surface; i.e., in any direction specified by a tangent vector. This amounts to a directional derivative, analogous to the directional derivatives applied to scalar-valued functions.

Differentiating the normal vector gives rise to something called the *shape operator*. It is defined as follows. Let $P_0 = \sigma(u_0, v_0)$ be a point on the surface Σ. Let $\mathbf{T}$ be a tangent vector to Σ at P_0. Let $\alpha(t) = \sigma(\mathbf{r}(t))$ be a curve on the surface having $\mathbf{T}$ as tangent vector at P_0; that is,

$$\mathbf{r}(0) = (u_0, v_0), \quad \alpha(0) = \sigma(\mathbf{r}(0)) = P_0, \quad \alpha'(0) = \mathbf{T}.$$

We set

$$S_{P_0}(\mathbf{T}) = -D_{\mathbf{T}}\mathbf{U} = -\frac{d}{dt}\mathbf{U}(\alpha(t))_{t=0}.$$

In fact, the value of $S_{P_0}(\mathbf{T})$ is independent of the choice of the curve α. (This is seen by an argument analogous to the one in Chapter 3.)

We call S_{P_0} an *operator* because it converts one tangent vector into another tangent vector. Indeed, since $\mathbf{U} \cdot \mathbf{U}$ is constant, the usual computation shows that $S_{P_0}(\mathbf{T}) \cdot \mathbf{U}(P_0) = 0$. So, $S_{P_0}(\mathbf{T})$ is perpendicular to $\mathbf{U}(P_0)$ and again lies in the tangent plane. The shape operator is easily checked to be *linear* on the two-dimensional tangent plane; that is, it satisfies

$$S_P(a\mathbf{T}_1 + b\mathbf{T}_2) = aS_P(\mathbf{T}_1) + bS_P(\mathbf{T}_2), \quad a, b \in \mathbf{R}.$$

Therefore, we can use the language of linear algebra to express it as a 2×2 matrix. If $\mathbf{T}_1$ and $\mathbf{T}_2$ are linearly independent vectors in the tangent plane, then

$$S_P(\mathbf{T}_1) = \alpha\mathbf{T}_1 + \beta\mathbf{T}_2, \quad S_P(\mathbf{T}_2) = \gamma\mathbf{T}_1 + \delta\mathbf{T}_2,$$

for some real numbers α, β, γ, δ. We say that the matrix $\begin{pmatrix} \alpha & \gamma \\ \beta & \delta \end{pmatrix}$ represents the operator S_P. (Of course, the four numbers depend on P; i.e., on the location on Σ, but we often suppress the subscript P.) The matrix representation is unique up to conjugation. In other words, if $\begin{pmatrix} \alpha' & \gamma' \\ \beta' & \delta' \end{pmatrix}$ results from another choice of basis, then $\begin{pmatrix} \alpha & \gamma \\ \beta & \delta \end{pmatrix} = A \begin{pmatrix} \alpha' & \gamma' \\ \beta' & \delta' \end{pmatrix} A^{-1}$ for some invertible matrix A. In fact, the invariants we will attach to the operator S_P will be unchanged by matrix conjugation (i.e., they are independent of the choice of basis).

Well, the actual computation of the matrix of S is "painful." Differentiating the unit normal and representing the result as a linear combination of σ_u and σ_v is a tedious chore. But it has been done many times by many people, and there is a standard presentation of the results, which we now supply. Define scalar-valued functions on the surface by

$$E = \sigma_u \cdot \sigma_u, \quad F = \sigma_u \cdot \sigma_v, \quad G = \sigma_v \cdot \sigma_v.$$

Also define

$$e = \mathbf{U} \cdot \boldsymbol{\sigma}_{uu}, \qquad f = \mathbf{U} \cdot \boldsymbol{\sigma}_{uv}, \qquad g = \mathbf{U} \cdot \boldsymbol{\sigma}_{vv}.$$

Then we have

$$S(\boldsymbol{\sigma}_u) = \frac{eG - fF}{EG - F^2}\boldsymbol{\sigma}_u + \frac{fE - eF}{EG - F^2}\boldsymbol{\sigma}_v,$$

$$S(\boldsymbol{\sigma}_v) = \frac{fG - gF}{EG - F^2}\boldsymbol{\sigma}_u + \frac{gE - fF}{EG - F^2}\boldsymbol{\sigma}_v.$$

(See *Modern Differential Geometry of Curves and Surfaces* by A. Gray (CRC Press, 1993), p. 275.) Now $\mathbf{U}$, being perpendicular to the tangent plane, satisfies $\mathbf{U} \cdot \boldsymbol{\sigma}_u = 0$. If we differentiate that equation with respect to v, we obtain $\mathbf{U}_v \cdot \boldsymbol{\sigma}_u + \mathbf{U} \cdot \boldsymbol{\sigma}_{uv} = 0$. But $\mathbf{U}_v = -S(\boldsymbol{\sigma}_v)$ by definition; hence

$$S(\boldsymbol{\sigma}_v) \cdot \boldsymbol{\sigma}_u = \mathbf{U} \cdot \boldsymbol{\sigma}_{uv} = f.$$

Reversing the roles of u and v, we see that $S(\boldsymbol{\sigma}_u) \cdot \boldsymbol{\sigma}_v$ also equals f; thus

$$S(\boldsymbol{\sigma}_u) \cdot \boldsymbol{\sigma}_v = \boldsymbol{\sigma}_u \cdot S(\boldsymbol{\sigma}_v).$$

Since $\boldsymbol{\sigma}_u$, $\boldsymbol{\sigma}_v$ span the tangent plane and since S is linear, we have $S(\mathbf{T}_1) \cdot \mathbf{T}_2 = \mathbf{T}_1 \cdot S(\mathbf{T}_2)$ for any two tangent vectors $\mathbf{T}_1$, $\mathbf{T}_2$. If we choose $\mathbf{T}_1$ and $\mathbf{T}_2$ orthonormal (i.e., of unit length and mutually perpendicular), and if $\begin{pmatrix} \alpha & \gamma \\ \beta & \delta \end{pmatrix}$ denotes the matrix of S with respect to this choice of basis, then $S(\mathbf{T}_1) \cdot \mathbf{T}_2 = \beta$ and $\mathbf{T}_1 \cdot S(\mathbf{T}_2) = \gamma$. Hence $\beta = \gamma$, and S is represented by a *symmetric* matrix (one which equals its own transpose). We use the fact from linear algebra that any symmetric matrix can be diagonalized. Thus, if $S = \begin{pmatrix} a & b \\ b & c \end{pmatrix}$, we can choose an orthogonal matrix $A = \begin{pmatrix} \cos\theta & \sin\theta \\ -\sin\theta & \cos\theta \end{pmatrix}$ so that $ASA^{-1} = \begin{pmatrix} k_1 & 0 \\ 0 & k_2 \end{pmatrix}$. In fact,

$$k_1, k_2 = \frac{(a + c) \pm \sqrt{(a - c)^2 + 4b^2}}{2}.$$

In linear algebra, the numbers k_1, k_2 are called the *eigenvalues* of the matrix. If we rotate the original basis by an angle θ, the new basis satisfies

$$S(\mathbf{T}_1) = k_1\mathbf{T}_1, \qquad S(\mathbf{T}_2) = k_2\mathbf{T}_2.$$

The perpendicular directions $\mathbf{T}_1$, $\mathbf{T}_2$ are called *principal directions* and the eigenvalues are called the *principal curvatures*. More generally, for any unit tangent vector $\mathbf{T}$, the number $k(\mathbf{T}) = S(\mathbf{T}) \cdot \mathbf{T}$ is called the *normal curvature* in the direction $\mathbf{T}$. Since any $\mathbf{T}$ may be written $\mathbf{T} = a\mathbf{T}_1 + b\mathbf{T}_2$, $a^2 + b^2 = 1$, then $k(\mathbf{T}) = a^2 k_1 + b^2 k_2$. Clearly (if $k_1 \geq k_2$), the normal curvature is largest (resp., smallest) in the principal direction of k_1 (resp., k_2).

EXERCISE. Show that the quadric surface $z = \frac{1}{2}(k_1 x^2 + k_2 y^2)$ has the same shape operator at the origin as Σ does at P if k_1 and k_2 are the principal curvatures there.

To clarify the exercise and the discussion that precedes it, we present two graphs. In both, the normal vector $\mathbf{U}$ and the principal directions $\mathbf{T}_1$, $\mathbf{T}_2$ are shown at one point on the surface. Curves through the point having those directions as tangent vectors are also displayed. In the left graph, the principal curvatures k_1, k_2 have opposite signs; in the right, the same sign. In the left graph, one principal direction is partly obscured by the surface.

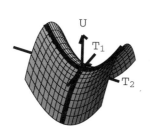
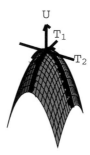

Let's give some geometric content to the notion of curvature. Let $\alpha(t) = \boldsymbol{\sigma}(\mathbf{r}(t))$ be a curve on Σ, $\alpha(0) = P$. Then $\alpha' \cdot \mathbf{U} = 0$. Differentiating again, we get $\alpha'' \cdot \mathbf{U} + \alpha' \cdot \mathbf{U}' = 0$, that is,

$$\alpha'' \cdot \mathbf{U} = \alpha' \cdot S(\alpha').$$

If α is unit speed and $\alpha'(0) = \mathbf{T}$, we can compute the normal curvature in the direction $\mathbf{T}$ as follows:

$$k(\mathbf{T}) = S(\mathbf{T}) \cdot \mathbf{T} = \alpha''(0) \cdot \mathbf{U}(P) = \kappa(0)\mathbf{N}(0) \cdot \mathbf{U}(P) = \kappa(0) \cos \theta,$$

where $\mathbf{N}(0)$ is the principal normal to the curve α at the point P, $\kappa(0)$ is the curvature of the curve there, and θ is the angle between the two normals $\mathbf{N}(0)$ and $\mathbf{U}(P)$. If we pick the curve to lie in the intersection of the surface with the plane determined by $\mathbf{U}(P)$ and $\mathbf{T}$, then $\mathbf{N}(0)$ must be in that plane; so it must be parallel to $\mathbf{U}(P)$ and $\cos \theta = \pm 1$. In other words, $k(\mathbf{T})$, the normal curvature of the surface in the direction $\mathbf{T}$, is ± 1 times the curvature of the curve $\kappa(0)$. (The ambiguity in sign is due to the fact that the surface only determines $\mathbf{U}$ up to sign; specifying $\mathbf{U}$ amounts to fixing an orientation.) If $k(\mathbf{T}) = \kappa(0) > 0$, the surface is bending toward $\mathbf{U}(P)$ in the direction $\mathbf{T}$; if $k(\mathbf{T}) < 0$, the surface is bending away in that direction. If $k(\mathbf{T}) = 0$, then $\kappa(0) = 0$ and $\mathbf{N}(0)$ is not defined—the surface is not bending at all in that direction.

We can enrich the discussion by introducing two more forms of curvature:

Definition. Define the *Gaussian curvature* K to be the product of the principal curvatures $K = k_1 k_2$, and the *mean curvature* H to be their average $H = \frac{1}{2}(k_1 + k_2)$.

REMARK. The Gaussian curvature is the *determinant* of the matrix S; and the principal curvature is one-half of its *trace*.

Experience has shown that the Gaussian curvature is the most useful of these invariants. The reason has to do with the following remarkable fact discovered by Gauss. Take three points P, Q, and R on the surface Σ that are close together, so that for all practical purposes

$$K(P) \approx K(Q) \approx K(R).$$

Join these points by a *geodesic triangle* in Σ. (The term "geodesic" comes from surveying; literally, it means "earth-dividing.") In other words, we join P to Q, P to R, and Q to R by curves lying in the surface, chosen to be as short as possible. (Line segments will usually not work, since the line segment in $\mathbf{R}^3$ joining two points in Σ may not be contained in Σ.) Since the triangle $\triangle PQR$ is "curved," the angles of this triangle need not add up exactly to π. However, Gauss observed that

$$\angle RPQ + \angle PQR + \angle QRP - \pi \approx \text{area}(\triangle PQR) \cdot K(P), \qquad \text{(Gauss)}$$

and that this formula is exact if the Gaussian curvature of Σ is constant.

The sphere of radius 1, which has constant Gaussian curvature, provides a good illustration of this phenomenon. Consider the geodesic triangle with vertices $P =$ the north pole and Q and R on the equator. Suppose the longitudes of the points Q and R differ by θ. The sides PQ and PR of the triangle are meridians, curves of constant longitude, which meet the equator at right angles. So

$$\angle RPQ = \theta, \qquad \angle PQR = \frac{\pi}{2}, \qquad \angle QRP = \frac{\pi}{2}.$$

This is illustrated in the following picture:

Geodesic Triangle on a Sphere

Thus the left-hand side of the equation (Gauss) is θ. On the other hand, the area of

$\triangle PQR$ is $\theta/(2\pi)$ times the area of the northern hemisphere, which is 2π, so

$$\text{area}\,(\triangle PQR) = \frac{\theta}{2\pi} \cdot (2\pi) = \theta,$$

as required by Gauss' formula.

Now return to our general discussion of curvature invariants. If $K(P) > 0$, the principal curvatures have the same sign, and the surface is bending in the same direction as you proceed in all directions from P. Locally, the surface looks like the paraboloid $2z = k_1 x^2 + k_2 y^2$ near P. If $K(P) < 0$, the principal curvatures have opposite sign. In one direction the surface bends "up," in the perpendicular direction it bends "down," and it could bend up and down in many different ways in between. (P is a *saddle point*, but the surface may resemble a hyperboloid, a monkey saddle, or some other wavy object). If $K(P) = 0$, there are two possibilities. If only one principal curvature vanishes, the surface resembles the trough-shaped paraboloid $2z = k_1 x^2$. If both vanish, then locally the surface is flat and we cannot really say much more. Geometers use the following terminology: points where $K > 0$ are called *elliptic*, points where $K < 0$ are called *hyperbolic*, and points where $K = 0$ are called *parabolic*.

Before proceeding, we redraw the two examples above, replacing the principal directions by the tangent plane. The tangent plane in the first example appears on both sides of the surface, because the Gaussian curvature is negative. In the second example, the Gaussian curvature is positive and the tangent plane lies completely on one side of the surface.

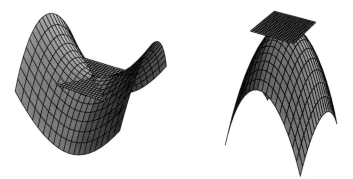

We shall not examine the mean curvature in detail in this chapter. Suffice it to say that a surface is called a *minimal surface* if its mean curvature vanishes. The exploration of minimal surfaces has a venerable history, because it has been known for a long time that minimal surfaces have an intuitively appealing application: a soap bubble tends to assume the shape of a minimal surface. We refer the interested reader to a course in differential geometry for more details. Instead, we simply present the formulas for K and H in terms of the shape operator invariants. Here they are, first for a general patch:

$$K = \frac{eg - f^2}{EG - F^2}, \qquad H = \frac{eG - 2fF + gE}{2(EG - F^2)},$$

and then for a Monge patch:

$$K = \frac{\varphi_{uu}\varphi_{vv} - \varphi_{uv}^2}{(1 + \varphi_u^2 + \varphi_v^2)^2}, \qquad H = \frac{(1 + \varphi_v^2)\varphi_{uu} - 2\varphi_u\varphi_v\varphi_{uv} + (1 + \varphi_u^2)\varphi_{vv}}{(1 + \varphi_u^2 + \varphi_v^2)^{3/2}}.$$

We conclude this chapter with a proposition that is reminiscent of the characterization results in Chapter 3, and with the analog of the congruence results (Theorem 4 and Corollary 5) in that chapter.

We say that a point P on Σ is *umbilic* if the normal curvature is constant there. Thus $k_1(P) = k_2(P)$, whence $K(P) \geq 0$.

Proposition. *If every point on a surface is umbilic, the surface is a portion of either a plane or a sphere.*

Proof. We show that not only is the normal curvature $k(\mathbf{T})$ independent of $\mathbf{T}$, but the value is independent of the point P. In fact if we differentiate the following equations with respect to v and u, respectively,

$$\mathbf{U}_u = -S(\boldsymbol{\sigma}_u) = -k\boldsymbol{\sigma}_u,$$
$$\mathbf{U}_v = -S(\boldsymbol{\sigma}_v) = -k\boldsymbol{\sigma}_v,$$

we obtain

$$\mathbf{U}_{uv} = \mathbf{U}_{vu} = -k_v\boldsymbol{\sigma}_u - k\boldsymbol{\sigma}_{uv} = -k_u\boldsymbol{\sigma}_v - k\boldsymbol{\sigma}_{vu}.$$

Therefore, $k_v\boldsymbol{\sigma}_u = k_u\boldsymbol{\sigma}_v$. Since $\boldsymbol{\sigma}_u$ and $\boldsymbol{\sigma}_v$ are linearly independent, $k_u = k_v = 0$. Hence, the principal (and normal) curvatures are constant. It follows that the Gaussian curvature is a nonnegative constant. If $K \equiv 0$, then $D_{\mathbf{T}}\mathbf{U} = 0$ for any direction $\mathbf{T}$. So $\mathbf{U}$ is constant and we have a plane. If $K > 0$, we reason as follows. Let P and Q be any two points on Σ. Select a curve α on Σ connecting the points, say $\alpha(0) = P$, $\alpha(1) = Q$. Define a function $\gamma(t) = \alpha(t) + (1/k)\mathbf{U}(\alpha(t))$. Differentiating, we find $\gamma'(t) = \alpha'(t) + (1/k)(\mathbf{U} \circ \alpha)'(t) = \alpha'(t) - (1/k)S(\alpha'(t)) = \alpha'(t) - \alpha'(t) = 0$. Thus γ is constant, say $\mathbf{c}$. Since the length of $\mathbf{U}$ is constantly equal to 1, we see that every point on the surface is at a constant distance $1/k$ from the point $\mathbf{c}$. ∎

Finally, here is the congruence theorem (see *Elementary Differential Geometry* by B. O'Neill (Academic Press, 1966), p. 299):

Theorem. *Let Σ_1, Σ_2 be two surfaces. Suppose there is a one-to-one map $\mathbf{F}$ from Σ_1 onto Σ_2 satisfying*

(i) **F** *takes the shape operator* S_1 *at every point* P_1 *on* Σ_1 *to the shape operator* S_2 *at the corresponding point* **F**(P_1) *on* Σ_2*; and*

(ii) **F** *preserves distances in the tangent planes,* $\|\mathbf{F}(\mathbf{T})\| = \|\mathbf{T}\|$.

Then there is a congruence of 3-space, composed of a rotation and a translation, that takes Σ_1 *onto* Σ_2.

A one-to-one correspondence between two surfaces satisfying property (ii) is the analog of a correspondence between two curves under which the speeds of the curves match up. Such a correspondence is called an *isometry*. Thus we see that isometric surfaces that have the same shape operator are congruent, giving substance to the assertion that *the shape operator encodes enough information to characterize the shapes of surfaces.*

This leads us to another question. When is there an isometry between surfaces that is *not* necessarily implemented by a congruence? An important aspect of this question was answered by Gauss, and again points to the fundamental role of the Gaussian curvature.

Theorem (Gauss). *Let* Σ_1, Σ_2 *be two surfaces. If* **F** *is a smooth map from* Σ_1 *to* Σ_2 *which is a (local) isometry, i.e., which preserves distances in the tangent planes, then* **F** *preserves Gaussian curvature, i.e., for every point* P_1 *on* Σ_1,

$$K_{\Sigma_1}(P_1) = K_{\Sigma_2}(\mathbf{F}(P_1)).$$

(However, **F** *need not preserve the mean curvature.)*

Curvature Calculations with *Mathematica*

Even though the calculations of the principal curvatures of a surface are quite messy to do by hand, it is not hard to automate the calculations using *Mathematica*. We develop *Mathematica* functions to do the calculations, and then apply them to some examples. The input for our functions should be a vector-valued function `surf` of two parameters `u` and `v`, giving the parametrization of our surface. We define the functions using delayed evaluation (`:=` in place of `=`), since we do not want to do the actual calculations before specifying the function `surf`. (See Question 7 in Chapter 10, *Mathematica Tips*, for an explanation.)

We start by computing the normal vector, unit normal vector, and the quantities e, E, etc. Since it is bad practice in *Mathematica* to use names beginning with a capital letter (as these could be confused with built-in functions and constants), we use `ee` to stand for E, etc.

```
In[15]:= <<Calculus'VectorAnalysis'
```

```
In[16]:= vecLength[vec_] := Sqrt[vec.vec]

In[17]:= normalVector[surf_, u_, v_] :=
        CrossProduct[D[surf, u], D[surf, v]] // Simplify

In[18]:= tangentPlane[surf_, u_, v_, x_, y_, z_] :=
        Simplify[({x, y, z} - surf) .
        normalVector[surf, u, v] == 0]

In[19]:= unitNorm[surf_, u_, v_] := normalVector[surf,u,v] /
        vecLength[normalVector[surf,u,v]] // Simplify

In[20]:= efgInvariants[surf_, u_, v_] :=
        {ee -> D[surf, u].D[surf, u],
        ff -> D[surf, u].D[surf, v],
        gg -> D[surf, v].D[surf, v],
        e -> unitNorm[surf, u, v].D[surf, {u, 2}],
        f -> unitNorm[surf, u, v].D[surf, u, v],
        g -> unitNorm[surf,u,v].D[surf, {v, 2}]}//Simplify

In[21]:= shapeOperator[surf_, u_, v_] := ((ee*gg-ff^2)^(-1)*
        {{e*gg - f*ff, f*ee - e*ff},
        {f*gg - g*ff, g*ee - f*ff}} /.
        efgInvariants[surf, u, v]) // Simplify
```

Now we can compute the principal curvatures, using *Mathematica*'s routine for computing the eigenvalues of a matrix.

```
In[22]:= principalCurvs[surf_, u_ ,v_] :=
        Eigenvalues[shapeOperator[surf, u, v]] // Simplify
```

We could compute the Gauss curvature and the mean curvature from the principal curvatures, but it is easier to compute them directly.

```
In[23]:= gaussCurv[surf_, u_, v_] := Det[
        shapeOperator[surf, u, v]] // Simplify

In[24]:= meanCurv[surf_, u_, v_] :=
        (shapeOperator[surf, u, v][[1]][[1]] +
        shapeOperator[surf, u, v][[2]][[2]])/2 // Simplify
```

Next we try our programs on three of the surfaces described at the beginning of the chapter. First the sphere:

```
In[25]:= principalCurvs[sphere, u, v]
```

$$\text{Out[25]= } \left\{ \mathrm{Csc}[v]\sqrt{\mathrm{Sin}[v]^2}, \mathrm{Csc}[v]\sqrt{\mathrm{Sin}[v]^2} \right\}$$

Both principal curvatures here are really 1, because $0 \leq v \leq \pi$, hence $\sin v \geq 0$ and $\sqrt{\sin^2 v} = \sin v$ cancels the cosecant. Therefore, the mean curvature and Gaussian curvature are everywhere equal to 1. As a check, we have

In[26]:= **gaussCurv[sphere,u,v]**

Out[26]= 1

In[27]:= **meanCurv[sphere,u,v]**

Out[27]= $\mathrm{Csc}[v]\sqrt{\mathrm{Sin}[v]^2}$

As before, this really equals 1. Next the hyperboloid of one sheet:

In[28]:= **principalCurvs[hyperboloid, u, v]**

Out[28]= $\left\{ \dfrac{1}{(1+2v^2)^{\frac{3}{2}}}, -\dfrac{1}{\sqrt{1+2v^2}} \right\}$

In[29]:= **gaussCurv[hyperboloid, u, v]**

Out[29]= $-\dfrac{1}{(1+2v^2)^2}$

The hyperboloid of one sheet has negative Gaussian curvature, with the curvature tending to zero at infinity. As a final example, we consider the helicoid.

In[30]:= **principalCurvs[helicoid, u, v]**

Out[30]= $\left\{ -\dfrac{1}{1+v^2}, \dfrac{1}{1+v^2} \right\}$

Since the principal curvatures are negatives of one another, the helicoid is an example of a minimal surface. Here's a check:

In[31]:= **meanCurv[helicoid, u, v]**

Out[31]= 0

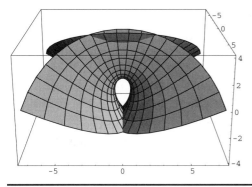

Problem Set F

SURFACES

When working on this problem set, you may use the programs in Chapter 6, *Geometry of Surfaces*. It is not necessary to retype them; they may be found on the disk accompanying this book.

1. Use the programs in Chapter 6 to compute the Gaussian and mean curvature for the remaining examples from the chapter: the sinusoidal cylinder, the cone, the monkey saddle, and the torus. In each case, interpret the results. Here are a few special points to look for:

   (a) In the case of the sinusoidal cylinder, what is the meaning of the Gaussian curvature calculation? (If you feel you need more information, compute the principal curvatures.)

   (b) In the case of the cone, what is the Gaussian curvature away from the vertex, and what does this mean? What happens at the vertex?

   (c) In the case of the monkey saddle, the surface does not have rotational symmetry. Nevertheless, what symmetry do you observe in the Gaussian curvature calculation?

   (d) In the case of the torus, what kind of symmetry is there and why? The Gaussian curvature varies from negative to zero to positive. Can you see why?

2. Compute formulas for the tangent plane, Gaussian curvature, and mean curvature at each point of the following Monge patch surfaces. Plot each surface. In each case, identify from your formula all points where the tangent plane is either horizontal or vertical, and reconcile this with your plot. Also, make any relevant comments about the Gaussian curvature—such as whether it's constant, where it is positive and where negative, and whether it manifests any symmetry properties (for instance, whether it is unchanged by translations in any direction, or rotations, etc.). Finally, state where the mean curvature vanishes.

   (a) The hyperbolic paraboloid: $z = xy$, $-2 \le x \le 2$, $-2 \le y \le 2$.

   (b) The paraboloid: $z = x^2 + y^2$, $-2 \le x \le 2$, $-2 \le y \le 2$.

   (c) The hemisphere: $z = \sqrt{1 - x^2 - y^2}$, $x^2 + y^2 \le 1$. (Hint: Since the function $\sqrt{1 - x^2 - y^2}$ is not defined on all of the square $-1 \le x, y \le 1$, for purposes of plotting you should replace it by the function `If[x^2+y^2 > 1, 0, Sqrt[1-x^2-y^2]]`.)

3. Compute a formula for the Gaussian curvature at each point of the following quadric surfaces. First parametrize and view the surface, as in the examples in Chapter 6, *Geometry of Surfaces*. (Select your `ViewPoint` carefully!) Discuss the sign of the Gaussian curvature and what it means. Point out places where the curvature is largest and smallest, any asymptotic behavior (i.e., what happens as you tend to infinity in some direction), and any symmetries.

   (a) The ellipsoid: $x^2 + y^2 + \dfrac{z^2}{4} = 1$.

   (b) The elliptic hyperboloid of one sheet: $x^2 + 2y^2 - z^2 = 1$.

   (c) The hyperboloid of two sheets: $x^2 + y^2 - \dfrac{z^2}{4} = -1$, $z > 0$.

4. Sometimes the formulas for Gaussian and mean curvature are complicated, and it is difficult to draw conclusions from them. In fact, even a *Mathematica* portrait of the surface may leave you a little puzzled about the way the surface is bending. In such a case it can be useful to have *Mathematica* plot the functions K and H, even if their formulas are "messy". This may help you describe the nature of the original surface. Illustrate this fact in the following cases by: plotting the surface, computing the invariants K and H, and then plotting K and H separately in three-dimensional plots. (Choose a consistent `ViewPoint` for the three plots in each case.) In some cases you might also want to use `ImplicitPlot` to display the shape of the curves along which K or H vanishes. For each surface, do your best to use the plots of K and H to say something meaningful about the shape of the original surface. Specifically, describe the elliptic and hyperbolic regions of the surface, any curves on which the surface is parabolic, where the

Gaussian curvature is "small", and where (in the elliptic regions) the surface is concave-up and where concave-down.

(a) A cubic surface: $z = x^3 - y^3$, $-3 \le x \le 3$, $-3 \le y \le 3$.

(b) A quartic surface: $z = (y - x^2)^2$, $-3 \le x \le 3$, $-3 \le y \le 3$.

(c) An exponential surface: $z = e^{x^2} - e^{-y^2}$, $-2 \le x \le 2$, $-1 \le y \le 5$.

5. The following surfaces have some sharp edges and vertices. Nevertheless, we can still use *Mathematica* to compute the curvature functions. Compute the Gaussian curvature for each surface, and then plot it as a function of the parameters. Observe where it is singular (i.e., where it is discontinuous, undefined, or blows up). Plot the surfaces and correlate the singular points you found to the plots. Is the curvature singular at *every* place on the surface where there is a sharp edge or vertex? Is the surface ever hyperbolic?

(Notes: The curvature computations for these surfaces may take a few minutes; be patient! For part (a), the `Simplify` commands in the programs for curvature calculation in Chapter 6 cause *Mathematica* to get hung up, so delete the word `Simplify` from the programs everywhere except at the end. Also, the option `PlotRange -> All` may be helpful for plotting the surface in (a).)

(a) The astroidal sphere:

$$\sigma(u, v) = (\cos^3 u \cos^3 v, \, \sin^3 u \cos^3 v, \, \sin^3 v).$$

(b) A figure eight surface:

$$\sigma(u, v) = (\cos u \cos v \sin v, \, \sin u \cos v \sin v, \, \sin v),$$

where in both examples the parameter ranges can be selected to be $0 \le u \le 2\pi$, $-\pi/2 \le v \le \pi/2$.

6. You know that the Gaussian and mean curvatures are defined in terms of the two principal curvatures by the formulas

$$K = k_1 k_2, \qquad H = (k_1 + k_2)/2.$$

(a) Solve these formulas for k_1 and k_2 in terms of K and H.

(b) What can you say about the relation between H^2 and K?

(c) Give conditions (in terms of H and K) for exactly one of the principal curvatures to vanish, and for both to vanish.

(d) Here's an application of the result in (c). Consider the surface $z = x^3 + x^2 y - 2x + y^2$. Compute H and K. The results, while complicated, are manageable. For example, show that K vanishes along a curve whose projection into the x-y plane is a parabola. By applying NSolve to the simultaneous equations given by the vanishing of the numerators of H and of K, show that all simultaneous solutions of $H = 0$ and $K = 0$ are complex. Thus, there is no point on the surface where both principal curvatures vanish. On the other hand, try to compute the principal curvatures directly. You will see that the formulas are so complicated that they are basically useless.

7. In this problem we study curvature for surfaces of revolution.

(a) For each of the following surfaces of revolution: plot the surface, compute its Gaussian curvature, and discuss the results. Use the examples in Chapter 6 as models.

(i) The hyperboloid of one sheet:

$$\sigma(u, v) = (\cosh u \cos v, \cosh u \sin v, \sinh u), \quad -1 \le u \le 1, \quad 0 \le v \le 2\pi.$$

(ii) A cubic surface:

$$\sigma(u, v) = (u \cos v, \, u \sin v, \, u^3 - 6u^2 + 9u - 2), \quad 0.25 \le u \le 3.75, 0 \le v \le 2\pi.$$

(iii) An inverted bell surface:

$$\sigma(u, v) = (u \cos v, u \sin v, 1 - e^{-u^2}), \quad 0 \le u \le 3, \quad 0 \le v \le 2\pi.$$

(b) Consider the surface of revolution determined by a profile curve $z = h(y)$ that does not intersect the z-axis:

$$\sigma(u, v) = (u \cos v, u \sin v, h(u)), \quad u \ge \delta > 0, \quad 0 \le v \le 2\pi.$$

Compute its Gaussian curvature. Based on the result and on examples (a(ii)) and (a(iii)), explain why there are no points of singularity for K (assuming h has a continuous second derivative). Determine where K is positive, negative, and zero, purely in terms of the geometry of the curve h. (Hint: The max/min and inflection points of h should play a role in your discussion. You may want to include a two-dimensional plot of the profile curves from (a(ii)) and (a(iii)).)

8. In this problem we study the notion of *umbilic* points. Recall that a point is called umbilic if its principal curvatures coincide there.

(a) Find a single equation relating K and H that must be satisfied at an umbilic point. Use it to explain why umbilic points can only occur where $K \ge 0$.

(b) At an umbilic point, the shape operator must be of the form $\begin{pmatrix} k & 0 \\ 0 & k \end{pmatrix}$, where k is the unique principal curvature. Why? (Hint: Recall from the discussion in Chapter 6 that the shape operator can be written in the form

$$\begin{pmatrix} \cos\theta & \sin\theta \\ -\sin\theta & \cos\theta \end{pmatrix} \begin{pmatrix} k_1 & 0 \\ 0 & k_2 \end{pmatrix} \begin{pmatrix} \cos\theta & -\sin\theta \\ \sin\theta & \cos\theta \end{pmatrix}$$

for some θ. See what happens to this formula when $k_1 = k_2 = k$.) Deduce that at an umbilic point, the unique principal curvature k satisfies

$$e = kE, \qquad f = kF, \qquad g = kG.$$

(Hint: Use `Solve`.)

(c) Use these facts to find all the umbilic points on the following surfaces:

(i) The saddle surface: $z = xy$.

(ii) The monkey saddle: $z = x^3 - 3xy^2$.

(iii) The elliptic paraboloid: $z = ax^2 + by^2$, $a, b > 0$.

(iv) The hyperbolic paraboloid: $z = ax^2 - by^2$, $a, b > 0$.

(Hint: You can use part (a) for (i) and (ii), but you will have to use part (b) to do (iii) and (iv). `Solve` is useful. In fact you can do both parts simultaneously because *Mathematica* will not distinguish the sign of the dummy variables you will use for a and b.

(d) Draw a picture of the elliptic paraboloid $z = 2x^2 + y^2$ with the umbilic points marked on it. `Point` can be used to create a point as a `Graphics3D` object, but set the `PointSize` so that it's easily visible. For the reasons explained in Problem 3 of Problem Set E, it helps to slide the surface down a bit so that it doesn't obscure the umbilic points.

9. In this problem we will study some surfaces with constant Gaussian curvature. We have already met several examples: a plane, a cone, a cylindrical surface, a sphere. In the first three $K \equiv 0$, the plane having both principal curvatures always vanishing, and the next two having exactly one principal curvature identically zero. The sphere has constant positive curvature. In this problem, we will find more exotic examples of surfaces of revolution with constant Gaussian curvature, both positive and negative, defined by means of *elliptic integrals*.

(a) Consider a surface of revolution:

$$\sigma(u, v) = (g(u)\cos v, g(u)\sin v, h(u)).$$

Compute its curvature K in terms of the functions g and h and their derivatives.

(b) Now assume the profile curve $(g(u), h(u))$ is of unit speed, in which case we have $(g')^2 + (h')^2 \equiv 1$. Solve for h', differentiate, and substitute into the expression for K obtained in (a) to show that $K = -g''/g$.

(c) Now to construct a surface of revolution of constant curvature 1, we solve the differential equation $1 = -g''(u)/g(u)$. Solve it, assuming for simplicity $g(0) = a > 0$, $g'(0) = 0$.

(d) The function h then is given by $h(u) = \int_0^u \sqrt{1 - g'(t)^2}\, dt$, which, given that you found the correct function g, is classically called an elliptic integral. Integrate to find the formula for h.

(e) Draw the three surfaces you get by choosing $a = 1$, $a = 0.5$, and $a = 1.5$. You should get a sphere, a football and a barrel, respectively. (You may use the interval $-\pi/2 \le u \le \pi/2$ for the first two plots, but you will have to shrink it for the third plot.)

(f) We can play a similar game to get a constantly negatively curved surface. Solve the differential equation $K = g''/g$ with the same initial data, except choose $a = 1$ for convenience. The function g will take a much simpler form when you simplify it using **ExpToTrig**. After obtaining this simplified form for g, compute h as in part (d). (The formula seems to involve imaginary numbers, but the values are actually real.) Draw the resulting surface. It is sometimes called the *bugle surface*. (Let u run from 0 to 1.5, and take the real part of h with **Re** to avoid error messages.)

10. In this problem we study minimal surfaces.

(a) Draw the helicoid

$$\sigma(u, v) = (u \cos v, u \sin v, v),$$

and show it is a minimal surface.

(b) Draw the catenoid (the surface of revolution spanned by the *catenary* $y = \cosh z$)

$$\sigma(u, v) = (\cosh u \cos v, \cosh u \sin v, u),$$

and show it is a minimal surface.

(c) There is a deformation of the catenoid into the helicoid through minimal surfaces. Consider the collection of surfaces, depending on a parameter t, defined by

$$\Theta(u, v, t) = \cos t(\sinh u \sin v, -\sinh u \cos v, v)$$
$$+ \sin t(\cosh u \cos v, \cosh u \sin v, u).$$

Check that for each t, $\Theta(u, v, t)$ parametrizes a minimal surface. Also check that this surface is a helicoid when $t = 0$ and a catenoid when $t = \pi/2$. Successively draw the surfaces corresponding to the values $t = 0$, $\pi/10$, $\pi/5$, $3\pi/10$, $2\pi/5$, $\pi/2$ and watch the helicoid turn into a catenoid. (Suggestion: Use the ranges $-2 \le u \le 2$, $0 \le v \le 2\pi$.) If you wish, you can animate the result by clicking on `Animate Selected Graphics` in the `Cell` menu at the top of your Notebook.

(d) The catenoid is a surface of revolution; the helicoid is not. In fact, the only surfaces of revolution that are minimal are catenoids. Prove this fact as follows. Consider the surface of revolution given by a curve $y = h(z)$, that is, the surface

$$\sigma(u, v) = (h(u) \cos v, h(u) \sin v, u).$$

Compute the mean curvature of the surface in terms of h and its derivatives. Set it equal to zero and solve the differential equation. You will see that the profile curve $z = h(u)$ is a catenary.

(e) Finally, here are two famous examples of minimal surfaces. Draw them and verify that their mean curvatures are identically zero.

(i) Scherk's minimal surface: $e^z \cos x = \cos y$: or

$$\sigma(u, v) = \left(u, v, \ln \frac{\cos v}{\cos u}\right),$$

where $-\pi/2 + 0.01 \le u \le \pi/2 - 0.01, -\pi/2 + 0.01 \le u \le \pi/2 - 0.01$.

(ii) Enneper's minimal surface:

$$\sigma(u, v) = \left(u - \frac{u^3}{3} + uv^2, v - \frac{v^3}{3} + u^2 v, u^2 - v^2\right),$$

where $-2 \le u \le 2, -2 \le v \le 2$.

Mathematica's default `ViewPoint` gives a nice picture in case (i). But try several choices (starting with the default) for a `ViewPoint` in case (ii), until you can clearly see the "hole."

Glossary of Some Useful *Mathematica* Objects

Commands

`Cross` The cross product of two vectors (not available in *Mathematica* 2.2).

`CrossProduct` The cross product of two vectors.

`CylindricalPlot3D` Plots surfaces given by equations in cylindrical coordinates.

`Det` Computes the determinant of a square matrix.

`DSolve` Symbolic differential equation solver.

`Eigenvalues` Computes the eigenvalues of a square matrix.

`ExpToTrig` Converts exponentials to trig or hyperbolic functions (not available in *Mathematica* 2.2).

`Graphics3D` Creates a graphics object in 3-space.

`If` Used for if/then/else conditional statements.

`ImplicitPlot` Plots the locus of an equation in 2-space.

`NSolve` Numerically solves a system of equations.

`ParametricPlot3D` Plots a curve or surface parametrically in 3-space.

`Plot3D` Plots a function of two variables in 3-space.

`Point` Creates a point as a `Graphics` or `Graphics3D` object.

`Show` Displays several graphics objects simultaneously.

`Solve` Solves a system of equations.

`SphericalPlot3D` Plots surfaces given by equations in spherical coordinates.

Options and Directives

`AspectRatio` Height to width ratio for graphics.

`AxesOrigin` Specifies where axes should intersect.

`PlotPoints` Specifies the number of points (in each direction) at which an expression is sampled in plotting.

`PlotRange` Specifies what portion of a plot should be shown. One possibility is `All`.

`PointSize` Specifies (relative) point size. Useful for `Point` or `ListPlot`.

`ViewPoint` Sets the viewpoint for three-dimensional graphics.

`WorkingPrecision` Sets the number of digits retained in numerical calculations. If round-off error seems to be a problem, try increasing this parameter.

Built-in Functions

`Im` Takes the imaginary part of a complex number.

`Re` Takes the real part of a complex number.

Packages

`Calculus`VectorAnalysis`` Needed for `CrossProduct`.

`Graphics`ImplicitPlot`` Needed for `ImplicitPlot`.

`Graphics`ʻ`ParametricPlot3D`ʼ Needed for `CylindricalPlot3D`, for an enhanced version of `ParametricPlot3D`, and for `SphericalPlot3D`.

`Graphics`ʻ`Shapes`ʼ Extra routines to construct standard shapes as `Graphics3D` objects.

`Graphics`ʻ`SurfaceOfRevolution`ʼ Extra routines to plot surfaces of revolution.

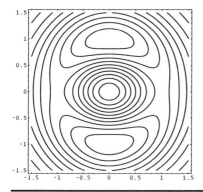

Chapter 7

OPTIMIZATION IN
SEVERAL VARIABLES

In this chapter, we will study generalizations to several variables of a few important applications of the derivative for a function f of one variable x. To make clearer the analogy between the one-variable and the several-variable cases, we begin with a quick review of the one-variable case, formulated in a way that will generalize easily.

The One-Variable Case

Suppose f is a *smooth* function defined on an interval of the real line. For our purposes, smooth will mean that f has a continuous second derivative. (The interval of definition may be finite, or it may be a half-line, or it may be the whole line.) We say f has a *local maximum* or *relative maximum* at $x = x_0$ if $f(x_0) \geq f(x)$ for all x close to x_0, and we say f has a *local minimum* or *relative minimum* at $x = x_0$ if $f(x_0) \leq f(x)$ for all x close to x_0. The word *extremum* means either a maximum or a minimum. Many real-world problems boil down to the mathematical problem of locating extrema. So how do we find the local extrema of a function f? There are two ways to do this:

(1) *analytically*, by an exact calculation from the formula for the function f; or
(2) *numerically*, by a numerical calculation from data associated with f. Often, numerical methods are augmented by a graphical procedure.

Calculus helps with both methods.

Analytic Methods

In pursuing an analytic method, we usually use one of the basic principles of one-variable calculus: *A (local) extremum always occurs at a critical point*, that is, at a point where the derivative f' of f vanishes. Actually, this is only a necessary condition: $f'(x_0)$ necessarily vanishes if $f(x)$ has a local extremum at $x = x_0$, but it can happen that a critical point is not a local extremum. (For example, think of the function $f(x) = x^3$ at $x = 0$.) Nevertheless, for the (smooth) functions we typically encounter, there are usually not many critical points, and it's often easy to tell by inspection which are local maxima, which are local minima, and which are neither. In cases where inspection doesn't work, there is always the *second derivative test*. If $f'(x_0) = 0$ and $f''(x_0) > 0$, then f has a local minimum at $x = x_0$, and if $f'(x_0) = 0$ and $f''(x_0) < 0$, then f has a local maximum at $x = x_0$. In fact, when $f'(x_0) = 0$ and $f''(x_0) > 0$, then f has a *strict local minimum* at $x = x_0$, in the sense that $f(x)$ is strictly bigger than $f(x_0)$ for x close to, but not equal to, x_0. Similarly, when $f'(x_0) = 0$ and $f''(x_0) < 0$, then f has a *strict local maximum* at $x = x_0$, in the sense that $f(x)$ is strictly smaller than $f(x_0)$ for x close to, but not equal to, x_0. Of course, it can happen that $f'(x_0) = 0$ and $f''(x_0) = 0$, but this behavior is *nongeneric*; i.e., it doesn't happen for typical functions. When $f'(x_0) = 0$ and $f''(x_0) = 0$, the second derivative test fails, but we can get more information from the third derivative, and so on.

Numerical Methods

Sometimes a function may not come with an explicit formula, or the formula may be too complicated to make it possible to solve directly for the critical points. In this case we can still locate local extrema numerically or graphically. Calculus can help us with this task in several ways. If the derivative f' is computable, but the equation $f'(x) = 0$ is too complicated to solve analytically, then we can use numerical root-finding procedures to locate numerically the solutions to $f'(x) = 0$. The most important such method, derived from calculus, is *Newton's method*. It is implemented in *Mathematica* via the `FindRoot` command. We will return to this in a moment.

Alternatively, we can use calculus to formulate an algorithm to hunt for local extrema, say local minima. This method is based on the familiar fact that when $f'(x_0) > 0$, then f is *increasing* at $x = x_0$, and when $f'(x_0) < 0$, then f is *decreasing* at $x = x_0$. Thus, suppose we start at some point $x = x_0$ where $f'(x_0) > 0$. (We might do this by explicit calculation of the derivative, by numerical estimation, or by looking at the graph of f.) Then f is *increasing* near $x = x_0$. So, if we want f to *decrease*, we have to move in the direction of *decreasing* x, that is, to the *left*. So we move some *step size* h to the left and look at $x_1 = x_0 - h$. If $f'(x_1) > 0$, then f is still increasing at x_1, so look at $x_2 = x_1 - h$, and so on. This process won't always converge (it certainly won't if f doesn't have a local minimum), but sometimes it will happen that eventually we come to a point x_n where $f'(x_n) > 0$ and $f'(x_{n+1}) < 0$ (with $x_{n+1} = x_n - h$). This means (since we are assuming f has a continuous derivative) that f' changes sign between $x = x_{n+1}$ and $x = x_n$. Hence, there is a local minimum between them. We can get a better approximation to this local minimum by repeating the same process with a smaller step size. Of course, if at some point x_i we have $f'(x_i) < 0$, then f is *decreasing* near $x = x_i$, so in this case we move to the *right*. We have just described a numerical algorithm for searching for a local minimum point. The algorithm has the advantage that it works even when we can't differentiate the formula for f analytically

A slightly fancier version of this procedure is implemented in *Mathematica* via the command `FindMinimum`. For example, the *Mathematica* command

```
FindMinimum[x^4 - 4x^2, {x, 1}]
```

searches for a minimum of the function $x^4 - 4x^2$, starting at $x = 1$. Let's try it.

```
In[1]:= FindMinimum[x^4 - 4x^2, {x, 1}]

Out[1]= {-4., x  →  1.41421}
```

This reflects the fact that $x^4 - 4x^2$ has a minimum value of -4 when $x = \sqrt{2} \approx 1.41421$. If instead we wanted to find a local *maximum* for f, we could simply apply the same algorithm to the function $-f$, since f has a local maximum where $-f$ has a local minimum.

Newton's Method

Newton's method is a process for solving nonlinear equations numerically. We might want to do this for reasons that have nothing to do with finding local extrema, but nevertheless the topics of equation-solving and *optimization* (looking for extrema) are closely linked. As we've already seen, looking for a local extremum of f forces us to try to solve the equation $f'(x) = 0$. We can also go in the other direction. If we want to solve an equation $g(x) = 0$, one way to do this is to look at the function $f(x) = g(x)^2$. The function f is nonnegative, so the smallest it could ever be is 0, and f takes the value 0 exactly where the equation $g(x) = 0$ is satisfied. So the solutions of $g(x) = 0$ occur at local minimum points of f.

To explain Newton's method for solving equations, let's take a simple example. Suppose we want to solve an equation such as $e^x = 3x$. We start by formulating an initial guess as to where a solution might be found. Since $e^1 = 2.7 \ldots$ and $3 \cdot 1 = 3$, it looks as if a good starting guess might be $x_0 = 1$. Next, we rewrite the equation in the form $g(x) = 0$. In this case we'd take $g(x) = e^x - 3x$. Near $x = x_0$, we have, by the *tangent line approximation*, the estimate $g(x) \approx g(x_0) + (x - x_0)g'(x_0)$. Setting this equal to zero, we get a linear equation for a (presumably) better approximation

x_1 to a solution. In other words, solving

$$g(x_0) + (x_1 - x_0)g'(x_0) = 0,$$

we get

$$x_1 = x_0 + (x_1 - x_0) = x_0 - \frac{g(x_0)}{g'(x_0)}.$$

Then we take x_1 as our new guess for a solution and compute an even better guess

$$x_2 = x_1 - \frac{g(x_1)}{g'(x_1)},$$

and so on. The process doesn't always converge, but it usually does if the starting value x_0 is close enough to a solution. In this case, the convergence to a true solution is usually quite fast. For example, in the given example, we have $g'(x) = e^x - 3$, so

$$x_1 = 1 - \frac{e - 3}{e - 3} = 1 - 1 = 0,$$

$$x_2 = 0 - \frac{1 - 0}{1 - 3} = \frac{1}{2},$$

$$x_3 = \frac{1}{2} - \frac{e^{.5} - 1.5}{e^{.5} - 3} = 0.61005965,$$

$$x_4 = 0.61005965 - \frac{e^{0.61005965} - 3 \cdot 0.61005965}{e^{0.61005965} - 3} = 0.61899678,$$

$$x_5 = 0.61899678 - \frac{e^{0.61899678} - 3 \cdot 0.61899678}{e^{0.61899678} - 3} = 0.61906128.$$

In fact, the process is automated in *Mathematica*'s `FindRoot` command.

```
In[2]:= FindRoot[Exp[x] == 3x, {x, 1}]
```

```
Out[2]= {x → 0.619061}
```

This tells us that, up to six-digit accuracy, the number 0.619061 satisfies the equation $e^x = 3x$.

One caution about the method: it tells you nothing about how many solutions there are. To find another solution (if there is one), you need to repeat the process with a different starting guess. For example, in this case, there is another solution, as you can see from

```
In[3]:= Plot[{Exp[x], 3x}, {x, 0, 2}];
```

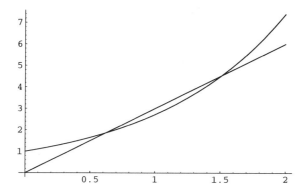

The graph suggests using $x = 1.5$ as a second starting point. And indeed

```
In[4]:=  FindRoot[Exp[x] == 3x, {x, 1.5}]

Out[4]=  {x  →  1.51213}
```

Note that the default in *Mathematica* is to give the answer to six significant digits. Actually, you can get any desired degree of precision by adjusting the options of `FindRoot`, notably `WorkingPrecision`. Here is an example:

```
In[5]:=  FindRoot[Exp[x] == 3x, {x, 1.5},
         WorkingPrecision -> 20]

Out[5]=  {x  →  1.5121345516578505078}
```

Functions of Two Variables

Now consider a function f of two variables x, y, defined in a suitable domain of the x-y plane. We assume our function is *smooth*. As with one variable, this means that f has continuous second partial derivatives. As before, we say f has a *local maximum* or *relative maximum* at (x_0, y_0) if $f(x_0, y_0) \geq f(x, y)$ for all (x, y) close to (x_0, y_0), and we say f has a *local minimum* or *relative minimum* at (x_0, y_0) if $f(x_0, y_0) \leq f(x, y)$ for all (x, y) close to (x_0, y_0). We say f has a *strict local maximum* or *strict relative maximum* at (x_0, y_0) if $f(x_0, y_0) > f(x, y)$ for all (x, y) close to, but not equal to, (x_0, y_0), and we say f has a *strict local minimum* or *strict relative minimum* at (x_0, y_0) if $f(x_0, y_0) < f(x, y)$ for all (x, y) close to, but not equal to, (x_0, y_0). Again, the word *extremum* means either a maximum or a minimum.

We still have the basic principle that *a (local) extremum always occurs at a critical point*, but now a critical point is a point where *both* partial derivatives vanish, in other words, where the *gradient vector* ∇f vanishes. (See Chapter 5.) In searching for local extrema, we can still proceed either analytically or numerically. Searching for analytic solutions is more complicated than in the one-variable case, since we need to solve two simultaneous equations in two unknowns. Sometimes the *Mathematica* routine `Solve` can help with the algebra. As in the one-variable case, it remains true that not every critical point is a local extremum. But it may be harder in the two-variable case to tell by inspection whether a critical point is a local maximum, a local minimum, or neither. Again, we can use a *second derivative test*.

Second Derivative Test

If $\nabla f(x_0, y_0) = 0$, so that (x_0, y_0) is a critical point for f, the second derivative test depends on the *Hessian matrix*

$$\text{Hess}\, f = \begin{pmatrix} \frac{\partial^2 f}{\partial x^2} & \frac{\partial^2 f}{\partial y \partial x} \\ \frac{\partial^2 f}{\partial x \partial y} & \frac{\partial^2 f}{\partial y^2} \end{pmatrix}.$$

The Hessian is a *symmetric matrix*; i.e., the entry in the upper right equals that in the lower left, since the two mixed second partial derivatives of a smooth function are equal to one another. The second derivative test is a bit complicated. It involves the determinant of a 2×2 matrix that you might have learned about in high school algebra:

$$\det \begin{pmatrix} a & b \\ c & d \end{pmatrix} = ad - bc.$$

The determinant of the Hessian matrix is called the *discriminant* of f. The test says:

(1) If $\nabla f(x_0, y_0) = 0$ and $\det \text{Hess}\, f > 0$ at (x_0, y_0), then f has a strict local extremum at (x_0, y_0). If $\partial^2 f/\partial x^2$ and $\partial^2 f/\partial y^2$ are both positive at (x_0, y_0), then f has a strict local *minimum* at (x_0, y_0). If $\partial^2 f/\partial x^2$ and $\partial^2 f/\partial y^2$ are both negative at (x_0, y_0), then f has a (strict) local *maximum* at (x_0, y_0). (Positivity of the determinant of a symmetric 2×2 matrix guarantees that the entries in the upper left and lower right corners are nonzero and have the same sign. Can you see why?)

(2) If $\nabla f(x_0, y_0) = 0$ and $\det \text{Hess}\, f < 0$ at (x_0, y_0), then f does *not* have a local extremum at (x_0, y_0). Instead we say that f has a *saddle point*. At a saddle point, f is increasing in some directions, and decreasing in others. The central point of a riding saddle has this property—whence the name. A nice way to visualize this behavior is with `ContourPlot`. At a saddle point, there will be two level curves intersecting as in the following example.

```
In[6]:=  ContourPlot[x^2 - y^2, {x, -3, 3}, {y, -3, 3},
         Contours -> { -3, -2, -1, 0, 1, 2, 3},
            ContourShading -> False];
```

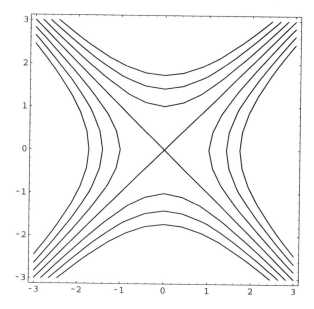

(3) When $\nabla f(x_0, y_0) = 0$ and $\det \operatorname{Hess} f = 0$, the second derivative test is inconclusive. There may or may not be a local extremum at (x_0, y_0), and if there is a local extremum, it may not be strict. Fortunately, it is still true that this behavior is *nongeneric*; i.e., it doesn't happen for "typical" functions.

Steepest Descent

Consider the algorithm we discussed above to locate a local minimum of a function f. In the one-variable case, the idea was to move in the direction in which the function is decreasing. In the two-variable case, however, there are infinitely many directions in which we can possibly move, not just two. The idea then is to move in the direction in which the function is decreasing the fastest. What direction is that? Well, recall from Chapter 5 that the directional derivative of f in the direction of a unit vector $\mathbf{u}$ is defined by

$$D_{\mathbf{u}} f(\mathbf{x}_0) = \lim_{t \to 0} \frac{f(\mathbf{x}_0 + t\mathbf{u}) - f(\mathbf{x}_0)}{t},$$

and is computed by the formula

$$D_{\mathbf{u}} f(\mathbf{x}_0) = \nabla f(\mathbf{x}_0) \cdot \mathbf{u}.$$

The directional derivative tells us how fast f is increasing in the direction of the unit vector $\mathbf{u}$. So to find the direction in which the function is decreasing the fastest, we need to make $D_{\mathbf{u}} f(\mathbf{x}_0)$ as negative as possible. If $\mathbf{u}$ is a unit vector, then

$$|D_{\mathbf{u}} f(\mathbf{x}_0)| = |\nabla f(\mathbf{x}_0) \cdot \mathbf{u}| \leq \|\nabla f(\mathbf{x}_0)\| \, \|\mathbf{u}\| = \|\nabla f(\mathbf{x}_0)\|,$$

so the minimum possible value of $D_{\mathbf{u}}f(\mathbf{x}_0)$ is precisely $-\|\nabla f(\mathbf{x}_0)\|$, and this minimum is achieved when $\mathbf{u}$ points in the direction of $-\nabla f(\mathbf{x}_0)$. Thus, we have arrived at the idea of the *method of steepest descent*: to locate a local minimum of a function of two (or more) variables, *move in the direction of the negative of the gradient*. With this modification, the method we outlined for the one-variable case works just fine. A version of it is implemented in the *Mathematica* routine `FindMinimum`. In fact, the command

```
FindMinimum[f[x, y], {x, a}, {y, b}]
```

searches for a local minimum of f starting from the point $\mathbf{x}_0 = (a, b)$. For example,

```
In[7]:=  f = x^4 - 4x*y + y^2;
```

```
In[8]:=  FindMinimum[f, {x, 1}, {y, 1}]
```

```
Out[8]=  {-4., {x  →  1.41421, y  →  2.82843}}
```

Mathematica searched for a minimum of the function $f(x, y) = x^4 - 4xy + y^2$, starting at $(1.0, 1.0)$, and returned the answer that a local minimum value of -4 is achieved at the point $x = 1.41421$, $y = 2.82843$. For this particular example, it is not so hard to see that there are three critical points, located at $(0, 0)$ and at $\pm(\sqrt{2}, 2\sqrt{2})$. We can better understand how steepest descent works for f by plotting the three critical points, some contour lines for f, and the gradient field of $-f$, using the methods of Chapter 5, *Directional Derivatives*. (The "minus sign" arises since it is the gradient of $-f$ that tells us in which direction f is decreasing the fastest.)

```
In[9]:=  <<Graphics`PlotField`
```

```
In[10]:= conPlot = ContourPlot[f, {x, -2, 2}, {y, -3, 3},
            ContourShading -> False, PlotPoints -> 100,
            Contours -> {-4, -2, 0, 2, 4, 10, 20}];
```

```
In[11]:= fieldPlot =  PlotGradientField[-f, {x, -2, 2},
            {y, -3, 3}, ScaleFunction -> (Tanh[#/5]&)];
```

```
In[12]:= critPtPlot =  ListPlot[{{-Sqrt[2], -2Sqrt[2]},
            {0,0}, {Sqrt[2], 2Sqrt[2]}},
            PlotStyle -> {PointSize[0.03]}];
```

```
In[13]:= Show[conPlot, fieldPlot, critPtPlot];
```

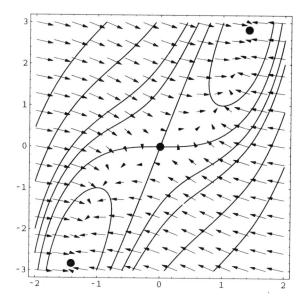

The steepest descent algorithm will follow the arrows, crossing a number of level curves of f, until it terminates at one of the critical points marked by a solid dot. Note that a saddle point is an *unstable* critical point for the method, in that roundoff errors will usually prevent landing on the saddle point unless we began there in the first place.

To search for a local maximum of f, apply the steepest descent algorithm to $-f$. Just as in the one-variable case, there is no guarantee that the steepest descent method will converge, and there is no way to be sure we have found *all* the local minima this way.

Multivariable Newton's Method

Since a single equation in two unknowns will usually have a whole curve of solutions (see the discussion of the Implicit Function Theorem in Chapter 6), the correct analog of the one-variable procedure is to solve not one equation but a system of two equations in two unknowns. We also need to do this to locate critical points of a function f, since they occur where both $\partial f/\partial x$ and $\partial f/\partial y$ are equal to 0. We use the language of vectors, suggested by the combination of $\partial f/\partial x$ and $\partial f/\partial y$ into a single vector function, ∇f. Let

$$\mathbf{f}(x,\, y) = (f_1(x,\, y),\, f_2(x,\, y)),$$

where f_1 and f_2 are both real-valued functions of x and y. Suppose we want to solve the vector equation

$$\mathbf{f}(x,\, y) = \mathbf{0}.$$

As before, we need an initial guess for a solution; call it $\mathbf{x}_0 = (x_0, y_0)$. In the one-variable Newton's method, we used the tangent line approximation to a function. In the two-variable case, we use the *tangent plane approximation*. Thus for x close to $\mathbf{x}_0$, we have

$$\mathbf{f}(\mathbf{x}) \approx \mathbf{f}(\mathbf{x}_0) + \big(\nabla f_1(\mathbf{x}_0) \cdot (\mathbf{x} - \mathbf{x}_0), \nabla f_2(\mathbf{x}_0) \cdot (\mathbf{x} - \mathbf{x}_0)\big).$$

Since we want to solve $\mathbf{f}(\mathbf{x}) = \mathbf{0}$, it is natural to try solving the approximation

$$\mathbf{f}(\mathbf{x}_0) + \big(\nabla f_1(\mathbf{x}_0) \cdot (\mathbf{x} - \mathbf{x}_0), \nabla f_2(\mathbf{x}_0) \cdot (\mathbf{x} - \mathbf{x}_0)\big) = \mathbf{0}.$$

In other words, we must solve the pair of simultaneous linear equations

$$\begin{cases} f_1(\mathbf{x}_0) + \nabla f_1(\mathbf{x}_0) \cdot (\mathbf{x} - \mathbf{x}_0) = 0, \\ f_2(\mathbf{x}_0) + \nabla f_2(\mathbf{x}_0) \cdot (\mathbf{x} - \mathbf{x}_0) = 0. \end{cases}$$

There is always the chance that this system might be *singular*, that is, it may not have a unique solution. But generically (i.e., if the starting value $\mathbf{x}_0$ and the function f are "typical"), this won't happen, so we get a unique solution $\mathbf{x}_1$. We use this as our next approximation to a solution and try solving

$$\begin{cases} f_1(\mathbf{x}_1) + \nabla f_1(\mathbf{x}_1) \cdot (\mathbf{x} - \mathbf{x}_1) = 0, \\ f_2(\mathbf{x}_1) + \nabla f_2(\mathbf{x}_1) \cdot (\mathbf{x} - \mathbf{x}_1) = 0, \end{cases}$$

to get $\mathbf{x}_2$, and so on. The process doesn't always converge, but when it does, the convergence is usually quite rapid. The *Mathematica* command

```
FindRoot[{f1[x, y] == 0, f2[x, y] == 0}, {x, a}, {y, b}]
```

implements a variant of this algorithm to search for a solution to the equations $f_1 = 0$, $f_2 = 0$, starting from the point $\mathbf{x}_0 = (a, b)$. For example,

```
In[14]:= FindRoot[{Exp[x] + Exp[y] == 6, x + y == 2},
         {x, 0}, {y, 1}]
```

```
Out[14]= {x → 0.548567, y → 1.45143}
```

Three or More Variables

Everything we've said also works for functions of three of more variables, with obvious modifications. The only thing that's a bit different is the second derivative test for critical points. If f is a function of three variables x, y, z, the local extrema still occur at critical points, that is, at points where the gradient

$$\nabla f = \left(\frac{\partial f}{\partial x}, \frac{\partial f}{\partial y}, \frac{\partial f}{\partial z}\right)$$

vanishes as a vector. To apply the second derivative test at such a critical point $x_0 = (x_0, y_0, z_0)$, we again look at the symmetric matrix of second partial derivatives

$$\text{Hess } f = \begin{pmatrix} \dfrac{\partial^2 f}{\partial x^2} & \dfrac{\partial^2 f}{\partial y \partial x} & \dfrac{\partial^2 f}{\partial z \partial x} \\ \dfrac{\partial^2 f}{\partial x \partial y} & \dfrac{\partial^2 f}{\partial y^2} & \dfrac{\partial^2 f}{\partial z \partial y} \\ \dfrac{\partial^2 f}{\partial x \partial z} & \dfrac{\partial^2 f}{\partial y \partial z} & \dfrac{\partial^2 f}{\partial z^2} \end{pmatrix}.$$

This time, the matrix contains not three, but six independent functions. (The three entries below the diagonal duplicate those above.) So the criteria for local maxima and minima are correspondingly more complicated than in the two-variable case. The best way to formulate the criteria is to say that x_0 gives a strict local minimum if the Hessian matrix is *positive definite* at x_0, and that x_0 gives a strict local maximum if the *negative* of the Hessian matrix is positive definite at x_0. In this formulation, the criteria even work for functions of four or more variables. The term positive definite is defined in linear algebra courses. For present purposes, we can use the following working definition: a symmetric $n \times n$ matrix is called *positive definite* if all its *principal minors* are positive. The principal minors are the determinants of the submatrices obtained by deleting the last i rows and i columns, for $i = n - 1, n - 2, \ldots, 0$. Thus the principal minors of the Hessian matrix of a function of three variables are the determinants

$$\dfrac{\partial^2 f}{\partial x^2}, \quad \det \begin{pmatrix} \dfrac{\partial^2 f}{\partial x^2} & \dfrac{\partial^2 f}{\partial y \partial x} \\ \dfrac{\partial^2 f}{\partial x \partial y} & \dfrac{\partial^2 f}{\partial y^2} \end{pmatrix}, \quad \det \begin{pmatrix} \dfrac{\partial^2 f}{\partial x^2} & \dfrac{\partial^2 f}{\partial y \partial x} & \dfrac{\partial^2 f}{\partial z \partial x} \\ \dfrac{\partial^2 f}{\partial x \partial y} & \dfrac{\partial^2 f}{\partial y^2} & \dfrac{\partial^2 f}{\partial z \partial y} \\ \dfrac{\partial^2 f}{\partial x \partial z} & \dfrac{\partial^2 f}{\partial y \partial z} & \dfrac{\partial^2 f}{\partial z^2} \end{pmatrix}.$$

When all three of these quantities are positive, we have a strict local minimum point. If the determinant of the Hessian is positive but either of the first two principal minors is nonpositive, then the critical point is definitely not a local minimum. (It will be a higher-dimensional saddle point.) If the determinant of the Hessian vanishes, the test is inconclusive. In general, the second derivative test for a local minimum of a function of n variables at a critical point involves positivity of n different functions of the second partial derivatives.

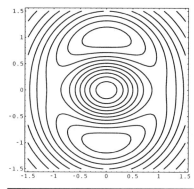

Problem Set G

OPTIMIZATION

1. Let $f(x, y) = x^4 - 3xy + 2y^2$.

(a) Compute the partial derivatives of f as well as its discriminant. Then use `Solve` to find the critical points and to classify each one as a local maximum, local minimum, or saddle point.

(b) Check your answer to (a) by showing that `FindMinimum` correctly locates the same local minima when you start at (0.5, 0.5) or at (−0.5, 0.5). Observe that in the second case, the numerical value of x appears to be slightly different from what you computed in (a). Why is this? Show that you can correct it by adjusting the `WorkingPrecision` option of the `FindMinimum` command.

(c) What happens when you apply `FindMinimum` with a starting value of (0, 0)? Explain your answer.

(d) What are the values of f at the extrema? Now, using `Plot3D`, graph the function on a rectangle that includes all the critical points. Use the graph and all the previous data to justify the assertion: "Sometimes symbolic and/or numerical computations are more revealing than graphical information".

2. Let $f(x, y) = (2x^2 + 3y^2)e^{-x^2 - y^2}$.

(a) Compute the partial derivatives of f as well as its discriminant. We would like to use `Solve` to find the critical points and to classify each one as a local

145

maximum, local minimum, or saddle point. But the complexity of the algebraic expressions for f_x, f_y, and Hess f give *Mathematica* some difficulty. Recompute these expressions by incorporating `Simplify`. Notice that there is a common exponential expression. Delete it before solving $f_x = f_y = 0$. Why is that legitimate? Then proceed with the classification.

(b) The point $(0, 0)$ gives a local minimum for f. Why is this obvious even without the second derivative test?

(c) When you apply `FindMinimum` to $-f$ with starting values of $(1, 0)$ and $(1, 0.00001)$, you get very different results. How can you explain this? (Hint: Draw a contour plot of f that encompasses the critical points $(0, 0)$ and $(0, 1)$. Draw another one encompassing all five critical points.)

3. Let $f(x, y) = x^6 + 3xy + y^2 + y^4$.

(a) Show that f remains unchanged if you replace x by $-x$ and y by $-y$. Hence, if (x, y) is a critical point of f, so is $(-x, -y)$. Thus, critical points other than $(0, 0)$ come in $\pm$ pairs.

(b) Compute the partial derivatives of f. Show that `Solve` fails to locate any of the critical points except for $(0, 0)$. Let's compensate numerically and then graphically. First show that `NSolve` succeeds. (Ignore the complex solutions.) Then confirm your findings by graphing the equations $f_x = 0$ and $f_y = 0$. You should see exactly one additional pair of critical points (in the sense of (a)). (Hint: Plot each of the equations separately and then combine the plots with `Show`. To plot each equation, you can either use `ImplicitPlot`, or else you can apply `ContourPlot` with the options `Contours -> {0}`, `ContourShading -> False`. If you wish to use `ImplicitPlot`, remember first to load the `Graphics'ImplicitPlot'` package.) Finally, compute the coordinates of the critical points (again numerically) by applying `FindRoot` to the equations for a critical point, using appropriate starting values estimated from your plot. Compare the numbers to those you obtained using `NSolve`. Classify the three critical points.

(c) Show that the *Mathematica* commands

```
FindMinimum[f[x, y], {x, 1}, {y, 1}]
FindMinimum[f[x, y], {x, 1.0001}, {y, 1.0001}]
```

both produce answers. But neither is a minimum. *Mathematica* senses something is wrong in the first instance, not in the second. Redo both computations, setting `WorkingPrecision -> 20` and `MaxIterations -> 50`. Explain what is happening. (Hint: Think about how the algorithm works in `FindMinimum`. What happens if the step size is too big?)

4. Find (numerically) all solutions of the system of nonlinear equations

$$\begin{cases} x^2 + y^4 = 2, \\ y = xe^{-x}. \end{cases}$$

First determine the number of solutions and their approximate locations graphically; then use `FindRoot` to compute them to six-decimal-place accuracy. (Hint: Adjustments to the options `WorkingPrecision` and/or `MaxIterations` may be necessary for `FindRoot` to get a sufficiently accurate answer.)

5. Use the second derivative test to find the conditions on a and b for the function $f(x, y, z) = x^2 + 3y^2 + z^2 + axy + byz + xz$ to have a strict local minimum at $(0, 0, 0)$. When is the second derivative test inconclusive? In fact the values of a and b for which f has a strict local minimum at the origin will be the interior of a solid region in the a-b plane. The second derivative test fails for points (a, b) on the boundary of that region. Your task is to find formulas in terms of a and b to describe the region. Finally, graph the region in the a-b plane where the conditions for a strict local minimum are satisfied.

6. The *Mathematica* routines `Solve` and `FindRoot` can also be applied to the system of Lagrange multiplier equations

$$\begin{cases} \nabla f(x, y) = \lambda \nabla g(x, y), \\ g(x, y) = 0 \end{cases}$$

for two functions f and g. Here f is the function to be maximized or minimized and g is the constraint function.

(a) Find (exactly, using Lagrange multipliers and `Solve`) the location and value of the absolute minimum of the function $xy - 3x$ on the circle $x^2 + y^2 = 1$. Give numerical values of your exact answers.

(b) Find (numerically, using Lagrange multipliers and `FindRoot`) the location and value of the absolute maximum and minimum of the function $e^x - \sin y$ on the ellipse $x^2 + 3y^2 = 1$. (Hint: You will have to do a `ContourPlot` or `ImplicitPlot` of two equations to locate some starting values for `FindRoot`. One equation is that of the ellipse. The other can be obtained by eliminating λ from the Lagrange equations. Why do you know $\lambda \neq 0$?) Check your answer by drawing a picture illustrating that the maximum and minimum occur where the level curves of $e^x - \sin y$ are tangent to the ellipse.

7. In practical optimization problems, we often have several constraints in the form of inequalities. Suppose for example that we want to find the absolute minimum of a function $f(x, y)$ subject to a constraint $g(x, y) \leq 0$. Geometrically, the constraint represents a region $\mathcal{R}$ in the x-y plane, with boundary defined

implicitly by the curve $g(x, y) = 0$. Two things can happen: either the minimum occurs at a critical point of f where $g(x, y) < 0$ (the case of an *interior minimum*) or else it occurs where $g(x, y) = 0$ (the case of a *minimum on the boundary*). We can use *Mathematica* first to find the interior minima (either by solving for the critical points or by using `FindMinimum`) and then to find the minimum on the boundary using Lagrange multipliers. The true global minimum is found by comparing the two answers and picking the point where f is the smallest. Carry out the complete procedure to find the location and value of the absolute minimum of the function $f(x, y) = 4 - x^2 - y^2 + x^4 + y^4$ subject to the constraint $g(x, y) = x^2 - 2x + y^2 - 3 \leq 0$. Be patient; there are many interior critical points and several very complicated solutions of the Lagrange multiplier equations. When you evaluate them numerically, you may use `Drop` to get rid of complex solutions of the equations.

8. This problem is a follow-up to the "baseball problem" (Problem 2) in Problem Set D. For simplicity we will neglect air resistance.

(a) In part (b) of the baseball problem, you determined the angle θ at which a batter should hit the ball to maximize the distance d traveled by the ball before it hits the ground, assuming that the initial speed v_0 at which the ball leaves the bat is held fixed. However, it is probably true that the batter can hit the ball with more power if his swing is level than if he needs to hit under the ball to loft it into the air. Take this into account by assuming that the speed with which the ball leaves the bat is not independent of θ but of the form $v_0/(1 + \sin \theta)$. Redo part (a) of the baseball problem with this variable initial velocity function (that depends on the angle); take $g = 32 \, \text{ft/sec}^2$, $h = 4 \, \text{ft}$, and $v_0 = 100 \, \text{ft/sec}$ (about 68 mi/hr). Use `FindMinimum` to locate the value of θ that maximizes the distance d. Does your answer seem reasonable?

(b) If a batter is trying for a single and not for a home run, the best strategy is probably not to maximize the distance traveled by the ball but rather to minimize the chance that it will be caught, subject to the constraint that the ball should be hit out of the infield. Thus the batter wants the ball to stay in the air as little time as possible. So solve the following constrained optimization problem: find the value of θ that minimizes the time t_0 till the ball hits the ground, subject to the constraint that $d \geq 120 \, \text{ft}$. Again take the speed of the ball to be $v_0/(1 + \sin \theta)$, but assume now that the batter can also adjust v_0 to be anything from 0 up to $120 \, \text{ft/sec}$. (You want to minimize t_0 as a function of v_0 and θ.)

9. Optimization of functions of several variables is exceptionally important in quantum mechanics. To demonstrate this, we consider for simplicity a particle of mass m free to move in one direction (say the x-axis), sitting in the "potential

well" described by a function $V(x)$ that tends to $+\infty$ as $x \to \pm\infty$. Then the "ground state" of the particle is given by the *wave function* ϕ which satisfies

$$\int_{-\infty}^{\infty} \phi(x)^2 \, dx = 1 \tag{*}$$

and for which the energy

$$E(\phi) = \int_{-\infty}^{\infty} \left(\frac{\hbar^2}{2m} \phi'(x)^2 + V(x)\phi(x)^2 \right) dx \tag{**}$$

is as small as possible. Here $\hbar$ denotes Planck's constant. (We will not attempt to justify this here, or to explain why such a function exists and is unique for reasonable choices of V.) Now, for any x, the size of $|\phi(x)|^2$ is roughly speaking a measure of how likely the particle is to be located close to x, and equation (*) says that the total probability of finding the particle *somewhere* is 1. In formula (**), the first term measures average "kinetic energy" and the second term measures average "potential energy."

A practical method that is often used to estimate the ground state energy and the form of the corresponding wave function is to guess some reasonable form for $\phi(x)$, depending on a few parameters, and then to minimize $E(\phi)$ as a function of these parameters. Here we experiment with this optimization technique in a few examples.

(a) For simplicity, take $m = 1$, $\hbar = 1$, and suppose $V(x) = x^2/2$, the case of a *harmonic oscillator*. In this case, it is known that the ground state wave function $\phi(x)$ is a *Gaussian function* of the form ce^{-ax^2} for some $a > 0$; where

$$c = \left(\int_{-\infty}^{\infty} e^{-2ax^2} \, dx \right)^{-1/2}$$

in order for (*) to be satisfied. Find the explicit formula for c in terms of a, and then compute the value of a that minimizes $E(\phi)$. What is the ground state energy? What fraction of this energy corresponds to kinetic energy and what fraction corresponds to potential energy? Plot the wave function ϕ.

(b) Again take $m = 1$ and $\hbar = 1$, but this time suppose $V(x) = x^4/2$, the case of the *anharmonic oscillator*. This potential is flatter than the previous one when $-1 < x < 1$, but it rises much faster for larger x. This time there is no closed-form expression for ϕ, so numerical methods are essential. By symmetry, we can see that ϕ must be an even function of x. Try taking $\phi(x) = ae^{-x^2}(1 + bx^2 + cx^4)$, where a is chosen in terms of b and c to satisfy (*). Find the formula for $E(\phi)$ as a function of b and c. (Hints: *Mathematica* has some trouble computing integrals of the form

$$\int_{-\infty}^{\infty} e^{-2x^2} p(x)^2 \, dx,$$

where $p(x)$ is a polynomial. You can help it by entering your integrals as

```
Integrate[Expand[Exp[-2*x^2]p[x]^2], x]
```

Use this syntax in your computations of a and of $E(\phi)$. Be patient; it takes *Mathematica* a while to compute these integrals.) Apply `FindMinimum` with starting values $b = 0$ and $c = 0$ to estimate the ground state energy, and plot the approximate wave function. What fraction of the ground state energy corresponds to kinetic energy and what fraction corresponds to potential energy? Compare with what happened in part (a). Do the results make physical sense?

(c) Again take $m = 1$ and $\hbar = 1$, but this time suppose $V(x) = \sec x$ for $|x| < \pi/2$, $V(x) = +\infty$ for $|x| \geq \pi/2$. Since the potential is infinite for $|x| \geq \pi/2$, the particle in this case is constrained to the interval $|x| < \pi/2$, and in formulas $(*)$ and $(**)$ you can replace the upper limit of integration by $\pi/2$ and the lower limit of integration by $-\pi/2$. This time try $\phi(x) = a\cos^2 x(1 + bx^2 + cx^4 + dx^6)$, where a is chosen in terms of b, c, and d to satisfy $(*)$. Apply `FindMinimum` with starting values $b = 0$, $c = 0$, $d = 0$ to estimate the ground state energy, and plot your approximate wave function. Compare with what happened in parts (a) and (b).

Glossary of Some Useful *Mathematica* Objects

Commands

`ContourPlot` Draws a contour plot of a function $f(x, y)$.

`D` Differentiation operator. One can also use an apostrophe, as in `f'[x]` for the derivative of `f[x]`.

`Drop` Drops one or more elements from a list.

`FindMinimum` Finds minimum values of a function of one or more variables.

`FindRoot` Finds numerical solutions of one or more transcendental equations.

`ImplicitPlot` Plots curves that are given implicitly as $f(x, y) = 0$.

`N` Evaluates an expression numerically.

`NSolve` Finds numerical solutions of an equation (or system of equations).

`Plot3D` Plots a surface $z = f(x, y)$.

`Show` Shows graphics; extremely useful for combining plots.

`Simplify` Simplifies complex algebraic expressions.

`Solve` Finds exact solutions of one or more algebraic equations.

Options and Directives

`Contours` Specifies number of or specific contours in `ContourPlot`.

`ContourShading` Controls shading in `ContourPlot`.

`DisplayFunction` Controls whether a graphics command displays its output; useful in suppressing extra output when using the `Show` command.

`MaxIterations` Controls the number of iterations *Mathematica* will allow when running a numerical routine (like `FindRoot`).

`Method` Selects the algorithm used in `FindMinimum`.

`WorkingPrecision` Specifies the precision in a numerical algorithm (such as employed in `FindRoot`).

Packages

`Graphics`Graphics`` Needed for `ImplicitPlot`.

`Miscellaneous`RealOnly`` Filters out nonreal solutions to equations.

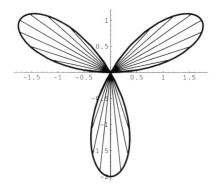

Chapter 8

MULTIPLE INTEGRALS

In this chapter, we explain how to use *Mathematica* to set up and to compute multiple integrals. We discuss techniques for visualizing regions in the plane and in space. We describe how to use these techniques to find the limits of integration in multiple integrals. Finally, we explain how to use these techniques in a variety of coordinate systems.

Automation and Integration

We believe that it is crucial for students to work problems by hand in order to master the methods of integration that form the backbone of a traditional multivariable calculus course. In addition, students must learn how to use technological tools like *Mathematica* to solve similar—but harder and more realistic—problems. We believe this dual strategy provides benefits in all of mathematics; indeed, in all of modern science.

It is commonly held that the advent of hand-held calculators has freed us from the drudgery of arithmetic. To some extent, this is true. We use calculators to balance checkbooks, to compute loan installments, and to free up time on mathematics and engineering examinations.

It is further held that there is no need to drill students in multiplication tables or similar techniques for arithmetic computation—the calculator will do it! Alas, there is a fatal flaw in this reasoning. If students cannot compute arithmetically, they are easily mislead by arithmetic mistakes and chicanery. If we have to multiply two three-digit numbers, we can expect the calculator to get it right. If our fingers slip, however, and the resulting answer consisting of a 12-digit number does not cause us instant suspicion, it is because our inability to compute has also left us unable to estimate. Without that ability, we will not know what mistakes—inadvertent or otherwise—are being foisted upon us.

Modern technology has raised the stakes. We now have tools (such as *Mathematica*) that automate the basic techniques of algebra (finding roots, multiplying matrices, simplifying radicals) and calculus (differentiating, integrating, finding limits). Just as we need to have mastered the basic rules of arithmetic in order to use a calculator effectively, so also must we master the basic rules of algebra and calculus in order to use *Mathematica* skillfully. If we do not understand the basic concepts, or have not mastered the basic ways to apply those concepts, then we cannot truly understand the output of a mathematical software system. Instead, one we are reduced to the meaningless manipulation of symbols. In fact, the situation is worse than that. Calculus was invented to give a mathematical formulation to the basic rules of the universe (the laws of gravity, the nature of chemical reactions, the rules of chance). Without understanding the mathematics, we have no hope of understanding the phenomena it is designed to represent.

The effective use of *Mathematica* for computing multiple integrals is an ideal venue in which to illustrate the importance of understanding concepts, as opposed to the mindless manipulation of symbols. In fact, *Mathematica* has already automated the computation of multiple integrals. In order to compute the double integral

$$\int_a^b \int_{f_1(x)}^{f_2(x)} g(x, y)\, dy\, dx,$$

you just enter the following command:

```
In[1]:= Integrate[g[x,y], {x, a, b}, {y, f1[x], f2[x]}]
```

$$\text{Out[1]}= \int_a^b \int_{f1[x]}^{f2[x]} g[x, y]\, dy\, dx$$

But that's the easy part. Once you know how to describe a region in the plane in terms of simple boundaries (the vertical line $x = a$, the vertical line $x = b$, and the graphs of two functions $y = f_1(x)$ and $y = f_2(x)$ which never cross), then computing the double integral is just two iterations of the process that you already know for computing an integral in one variable. The hard part is figuring out how to describe a complicated region in such a simple fashion.

Regions in the Plane

We want to look at the following problem, which is an archetype of an entire class of problems found in every textbook ever written about multivariable calculus.

Find the area between the two curves

$$y = \sin(x+1) \quad \text{and} \quad y = x^3 - 3x + 1.$$

We start by defining the two curves as *Mathematica* expressions, and plotting them on the same graph.

```
In[2]:=  f = Sin[x + 1];
         g = x^3 - 3x + 1;

In[3]:=  twocurves = Plot[{f, g}, {x, -3, 3},
            PlotStyle -> {{}, Thickness[0.008]}];
```

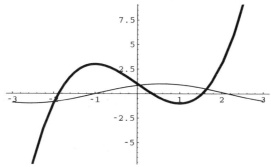

We drew the polynomial with a thicker style so we can distinguish the two curves; we also gave the graph a name so we can reuse it.

The graph gives us several pieces of information. First of all, there are three points of intersection. In order from left to right, let's call the x-coordinates x_1, x_2, and x_3. The region between the two curves consists of two distinct pieces, one above the interval $[x_1, x_2]$ and the other above $[x_2, x_3]$. Moreover, the two curves cross at x_2, exchanging the roles of top and bottom curve. Thus, when we finally get around to computing an integral, we'll need to compute the contribution of each of these regions separately.

The `Solve` command can't find the points of intersection in this case, because of the transcendental function. Instead, we'll use `FindRoot`.

```
In[4]:=  x1 = x /. FindRoot[f == g, {x, -2}]
         x2 = x /. FindRoot[f == g, {x, 0}]
         x3 = x /. FindRoot[f == g, {x, 2}]
```

```
Out[4]=  -1.98099
```

```
Out[5]=  0.0450476
```

```
Out[6]=  1.6383
```

In order to identify the region visually, we are going to "shade" it with vertical lines. We'll need the *Mathematica* commands `Graphics` (to represent basic graphics elements) and `Line` (to represent the particular graphics element known as a line). The following command draws several lines, each of which connects a point $(x, f(x))$ on one curve via a vertical line to a point $(x, g(x))$ on the other curve.

```
In[7]:=  Show[twocurves, Table[Graphics[Line[
             {{x, f}, {x, g}}]], {x, x1, x3, 0.2}]]]
```

With the graph to guide us, we can now confidently compute the area using double integrals.

```
In[8]:=  leftarea = Integrate[1, {x, x1, x2}, {y, f, g}]
```

Out[8]= 4.00506

```
In[9]:=  rightarea = Integrate[1, {x, x2, x3}, {y, g, f}]
```

Out[9]= 2.0066

```
In[10]:= area = leftarea + rightarea
```

Out[10]= 6.01167

EXERCISE. We were careful to break the region into two pieces; that operation is necessary to solve the more general problem of integrating any function of two variables over a region. When you only want to compute an area, you can sometimes get a quicker estimate by integrating the absolute value of the difference between the bounding functions. Try the command

```
NIntegrate[Abs[f - g], {x, x1, x3}]
```

and compare its result with the answer we just computed.

Viewing Simple Regions

A region is called *vertically simple* if each vertical line intersects the region in at most one connected segment. We can automate the technique we just used to view a vertically simple region by writing our own *Mathematica* command. Recall that we drew two curves, one with a normal `PlotStyle` and the other with a thicker style; then we drew several vertical lines connecting the two curves. We can write a command to carry out the same two steps for any vertically simple region. The arguments that we will need to supply to the command come from those we needed to supply to the **Integrate** command: the left boundary constant **a**, the right boundary constant **b**, the bottom bounding function **f**, and the top bounding function **g**.

```
In[11]:= verticalRegion[a_, b_, f_, g_] :=
        Module[{twocurves},
          twocurves = Plot[{f, g}, {x, a, b},
            PlotStyle -> {{}, Thickness[0.008]}];
          Show[twocurves, Table[Graphics[Line[
            {{x, f}, {x, g}}]], {x, a, b, (b-a)/20}]]]
```

The definition of **verticalRegion** wraps everything inside a **Module** command. This construction serves two purposes. First, it allows us to group several statements into a single structure. Second, it allows us to use local variable names (in this case, **twocurves**) without having to worry about whether those names have been used elsewhere in the same *Mathematica* session.

We could reproduce our previous picture by typing

```
verticalRegion[x1, x3, f, g]
```

As that picture indicates, we can only refer to **f** as the bottom and **g** as the top if we were careful enough to divide the region up into pieces before trying to graph it. A visual inspection of the graph will tell us if we got it right; the top curve should be drawn with a thicker style than the bottom curve. (On a color monitor, you might want to use **RGBColor** to define distinct plot styles instead of **Thickness**.)

A region is called *horizontally simple* if each horizontal line intersects the region in at most one connected segment. In Problem 3 of Problem Set H, you will be asked to modify the **verticalRegion** program to draw horizontally simple regions. The following graph shows a horizontally simple region between the curves $x = \sin(y)$ and $x = \cos(y)$ that was drawn with such a program.

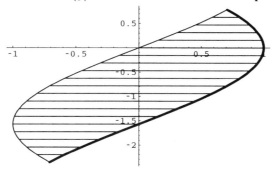

We can find the area of this region by integrating first with respect to x, and then with respect to y. Recall that the `Integrate` command requires you to enter the variables in the order that the limits appear on the integral signs, which is the opposite of the order of evaluation.

In[12]:= **Integrate[1, {y, -3Pi/4, Pi/4}, {x, Sin[y], Cos[y]}]**

Out[12]= $2\sqrt{2}$

Polar Regions

Many regions are more naturally described using polar coordinates instead of rectangular coordinates. Using a different coordinate system provides us with different natural ways to chop up regions. As an example, consider the following region R. Start with a circle of radius 2. Now remove a circle of radius 1 that is tangent to the first circle. What is the area of the region R that remains?

We don't need calculus to answer this question. The outer circle has area 4π; the deleted circle (regardless of its position inside the bigger circle) has area π, so the remaining area is just 3π. So why would we go to the trouble of carefully graphing the region and evaluating a double integral in polar coordinates just to find out something we already know? Well, consider the following picture:

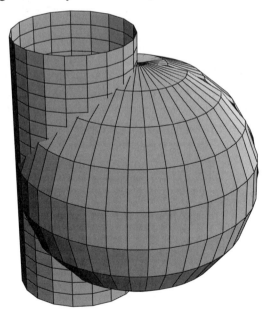

This graph shows a sphere of radius 2 intersected by a cylinder of radius 1; one edge of the cylinder passes through the center of the sphere, and the other edge is tangent to the sphere at the equator. (Note: The curve along which these surfaces intersect is called Viviani's curve, and was studied in Chapter 2.) Suppose we drill a hole where the cylinder passes through the sphere. What is the volume of the portion of the sphere that remains? Finding the volume of this solid requires us to integrate a function over the plane region R—but this time, we don't know the answer in advance. We will return to this problem later.

We are going to define a *Mathematica* command that plots polar regions in the same way `verticalRegion` plots vertically simple regions. This command will plot regions corresponding to a polar integral of the form

$$\int_\alpha^\beta \int_{r_1(\theta)}^{r_2(\theta)} f(r, \theta)\, r\, dr\, d\theta.$$

In other words, we want to be able to display regions defined in polar coordinates that are bounded by the lines $\theta = \alpha$ and $\theta = \beta$, and lying between the curves $r = r_1(\theta)$ and $r = r_2(\theta)$.

In order to draw lines on our graphs, it is useful to be able to convert from polar coordinates to rectangular coordinates. The following commands will do that:

```
In[13]:= xp[r_] := r*Cos[t]
         yp[r_] := r*Sin[t]
```

Now we can define a command that plots regions in polar coordinates. The command to plot functions in polar coordinates lives in the `Graphics` package, so we start by loading it. Then we modify the `verticalRegion` program to use `PolarPlot` instead of `Plot`.

```
In[14]:= <<Graphics`Graphics`

In[15]:= polarRegion[a_, b_, R1_, R2_] :=Module[{twocurves},
           twocurves = PolarPlot[{R1, R2}, {t, a, b},
             PlotStyle -> {{}, Thickness[0.008]}];
           Show[twocurves, Table[Graphics[Line[
             {{xp[R1], yp[R1]}, {xp[R2], yp[R2]}}]]],
             {t, a, b, (b-a)/30}]]]
```

The `polarRegion` command, like the `verticalRegion` command developed earlier, actually produces two graphs; one without the lines and one with the lines. You can modify the command using `DisplayFunction` to suppress the first graph automatically. The Glossary contains an example showing you how to do this using `DisplayFunction`. In the example below, we will only show the second graph.

Here is the graph of the region R between the two circles.

```
In[16]:= circ1 = 2 Cos[t];
         circ2 = 4 Cos[t];

In[17]:= polarRegion[-Pi/2, Pi/2, circ1, circ2];
```

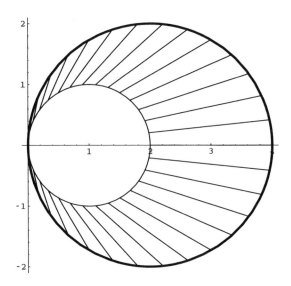

Now let's check that we can compute the area correctly.

`In[18]:= Integrate[r, {t, -Pi/2, Pi/2}, {r, circl, circ2}]`

`Out[18]=` 3π

Next, we return to the problem of computing the volume of the part of the sphere that remains after the intersection with the cylinder is removed. Observe first that the sphere of radius 2 with center located at $(x, y, z) = (2, 0, 0)$ has the rectangular equation

`In[19]:= rectSphere = ((x - 2)^2 + y^2 + z^2 == 4);`

To convert to a polar equation, we write

`In[20]:= polarSphere = rectSphere /.`
`        {x -> r*Cos[t], y -> r*Sin[t]} // Simplify`

`Out[20]=` $r^2 + z^2 - 4r\,\mathrm{Cos}[t] == 0$

Solving this equation for z, we easily see that $z = \pm\sqrt{4r\cos(t) - r^2}$. Using the symmetry between the top half and the bottom half of the solid, we see that the desired volume should be

`In[21]:= 2*Integrate[r*Sqrt[4r*Cos[t] - r^2],`
`        {t, -Pi/2, Pi/2}, {r, circl, circ2}]`

`        Integrate::gener: Unable to check convergence`

`Out[21]=` $\dfrac{16}{9}(4 + 3\pi)$

In spite of the warning message, *Mathematica* managed to compute the integral correctly. The warning arises because of the possibility that the term under the square root could be negative. You can check the answer by computing the integral in smaller steps: First, compute the indefinite integral with respect to r, and then differentiate to verify that it is correct. Then evaluate at the endpoints, where $r = 2\cos(t)$ and $r = 4\cos(t)$. Next, use the remaining symmetry in the problem and integrate with respect to t over the interval from $-\pi/2$ to $\pi/2$.

Viewing Solid Regions

Let's turn our attention now to solid regions. Our first task is to devise an analog of the **verticalRegion** command that will allow us to visualize three-dimensional solid regions. We will need to supply more arguments; we will need to replace **Plot** with **ParametricPlot3D**; and we will need to find a way to draw lines connecting the top and bounding surfaces. Here is a command that meets our requirements:

```
In[22]:= viewSolid[{x_,a_,b_}, {y_,f_,g_}, {z_,F_,G_}] :=
        Module[{u, v, xx, yy, surfs, pplot},
           xx = a + u*(b - a);
           yy = f + v*(g - f) /. x -> xx;
           surfs = {{x, y, F}, {x, y, G}} /. {x->xx, y->yy};
           pplot = ParametricPlot3D[Evaluate[surfs],
             {u, 0, 1}, {v, 0, 1},
             DisplayFunction -> Identity];
           Show[pplot,
             Table[Graphics3D[Line[surfs]],
                {u, 0, 1, 0.2}, {v, 0, 1}],
             Table[Graphics3D[Line[surfs]],
                {u, 0, 1}, {v, 0, 1, 0.2}],
             DisplayFunction -> $DisplayFunction]]
```

Since this is a lot of code to digest at one time, let's go through it step by step. The first line simply describes the arguments to **viewSolid**. These are compatible with the arguments that you need to supply to the **Integrate** command to compute the triple integral

$$\int_a^b \int_{f(x)}^{g(x)} \int_{F(x,y)}^{G(x,y)} h(x,y,z)\, dz\, dy\, dx.$$

(Of course, the `Integrate` command also requires the integrand at the beginning of the argument list.)

You should note that `viewSolid` differs from `verticalRegion` by requiring you to specify the variables x, y, and z as arguments to the command. In the plane, we only had two choices for the order of integration, and only two kinds of regions to display. So, it made sense to construct separate commands. For `verticalRegion`, we could assume that y ranged between functions of x while x ranged between constants; `horizontalRegion` reversed the role of the variables. In three dimensions, we have six possible orders of integration, and six kinds of regions to display. In this case, it is easier to create a single command that will work for all six possible orders. We will illustrate the use of `viewSolid` with integration order z-y-x. So, the parameters a and b should be given constant values a and b; the parameters f and g should be expressions representing the mathematical functions $f(x)$ and $g(x)$, and the parameters F and G should be expressions representing the mathematical functions $F(x, y)$ and $G(x, y)$. When using a different order, you will have to change the variables in the functions accordingly, and remember that "up" in your picture is not necessarily the z-direction.

The second line of code declares that this function is a `Module`, and it lists the local variables that will be used. The local variables u and v are the parameters used in the call to `ParametricPlot3D`. The next two lines assign values to the local variables xx and yy. As u ranges from 0 to 1, the local variable xx ranges from a to b. As v ranges from 0 to 1 for fixed u, the local variable yy ranges from $f(x)$ to $g(x)$.

The next line of code uses xx and yy to rewrite the surfaces $z = F(x, y)$ and $z = G(x, y)$ as parametric surfaces in terms of the parameters u and v. The following three lines construct a three-dimensional parametric plot of those surfaces, without displaying the plot. The final section of the code combines the parametric plot with two sets of lines. The first `Table` draws lines at the extreme v edges of the graph; the second `Table` does the same thing at the extreme u edges.

Here is an example. We will view the solid region bounded below by the x-y plane, bounded above by the plane $z = (x + y)/4$, and lying over the plane region bounded by $x = 1$, $x = 2$, $y = x/2$, and $y = x$.

```
In[23]:= viewSolid[{x, 1, 2}, {y, x/2, x}, {z, 0, (x+y)/4}];
```

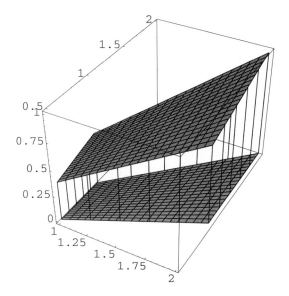

Now let's look at a more interesting example: we want to compute the volume of the solid region R bounded by the parabolic cylinder $z = 4 - y^2$ and the paraboloid $z = x^2 + 3y^2$. The first thing to do is to determine where the surfaces intersect. Let's start by telling *Mathematica* about the equations.

```
In[24]:= cyl = (z == 4 - y^2);
         par = (z == x^2 + 3y^2);
```

We can eliminate the common variable z from the equations to determine the projection of the curve of intersection onto the x-y plane.

```
In[25]:= xycurve = Eliminate[{cyl, par}, z]
```

Out[25]= $x^2 == 4 - 4y^2$

As is typically the case, the result is an equation that implicitly defines a curve in the x-y plane. We can use the **ImplicitPlot** command to see the curve, which we recognize as an ellipse.

```
In[26]:= <<Graphics`ImplicitPlot`
```

```
In[27]:= ImplicitPlot[xycurve, {x, -2, 2}, {y, -1, 1}];
```

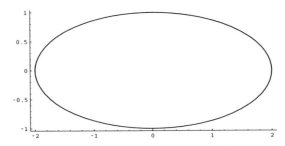

Now we can use the `Solve` command to get the limits for y as a function of x.

```
In[28]:= ysols = Solve[xycurve, y]
```
$$\text{Out[28]= } \left\{\left\{y \to -\frac{1}{2}\sqrt{4-x^2}\right\}, \left\{y \to \frac{\sqrt{4-x^2}}{2}\right\}\right\}$$

```
In[29]:= ylim = y /. Last[ysols]
```
$$\text{Out[29]= } \frac{\sqrt{4-x^2}}{2}$$

Before viewing the solid, we need to think for a moment about whether the parabolic cylinder $z = 4 - y^2$ or the paraboloid $z = x^2 + 3y^2$ is on the top. Since the point $(0,0)$ lies inside the ellipse in the x-y plane, we can do a simple hand computation at this point to see that the height of the paraboloid is 4, while the height of the parabolic cylinder is 0. So, the paraboloid is on top, and the appropriate command to view the solid is as follows:

```
In[30]:= viewSolid[{x, -2, 2}, {y, -ylim, ylim},
          {z, x^2 + 3y^2, 4 - y^2}];
```

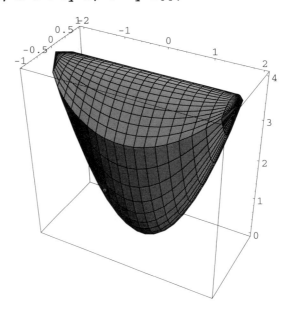

Now we can integrate to find the volume.

```
In[31]:= Integrate[1, {x, -2, 2}, {y, -ylim, ylim},
         {z, x^2 + 3y^2, 4 - y^2}];
```

```
Out[31]= 4π
```

A More Complicated Example

In this section, we want to compute the volume of the intersection of two solids, a torus and a cylinder:

A torus is the result of revolving a circle around a disjoint axis. Here we take a circle of radius 1 and revolve it around an axis that lies four units from the center of the circle. Taking the axis of revolution to be the z-axis, and parametrizing the circle by $(4 + \cos u, 0, \sin u)$ for $0 \le u \le 2\pi$, we find that the torus is defined parametrically in *Mathematica* by

```
In[32]:= torus = {(4 + Cos[u])Cos[v], (4+Cos[u])Sin[v],
         Sin[u]};
```

where $0 \le u \le 2\pi$ and $0 \le v \le 2\pi$. Writing $r = \sqrt{x^2 + y^2}$, we have

$$r^2 = x^2 + y^2 = (4 + \cos u)^2$$

and $z = \sin u$. Thus, the torus is defined in rectangular coordinates by:

```
In[33]:= toreqn = ((r-4)^2 + z^2 == 1) /.
         {r -> Sqrt[x^2 + y^2]} // Simplify
```

$$Out[33]= \left(-4 + \sqrt{x^2 + y^2}\right)^2 + z^2 == 1$$

We intersected the torus with a cylinder of radius 1 that passes through the center of the torus. The cylinder is defined parametrically by

```
In[34]:= cylinder = {x, Cos[t], Sin[t]};
```

and it is defined in rectangular coordinates by the following equation:

```
In[35]:= cyleqn = (y^2 + z^2 == 1);
```

The task before us is to write down a triple integral that computes the volume of the solid obtained by intersecting the torus and the cylinder. The hardest part of this task is choosing the order (and finding the limits) of integration. We plan to create a tool for visualizing the solid that will help us make that choice.

Here's an idea for doing that. Suppose, for example, that we replace x by a constant value in the equations that define the cylinder and the torus. This algebraic substitution corresponds to the geometric operation of intersecting both solids with a plane. In this case, the plane lies at right angles to the cylinder, so it intersects it in a circle. The intersections with the torus will vary. By plotting the intersections for various values of x, we can get a good idea of the different cross sections of the intersection of the two solids. If the cross sections are simple, then x is a good choice for the final variable in the order of integration. If the cross sections are complicated, then we should try again with a different variable instead of x. The following command will draw the cross sections:

```
In[36]:= Table[ImplicitPlot[
         Evaluate[{toreqn, cyleqn, y^2 + z^2 == 25}],
         {y, -5, 5}, {z, -5, 5}, PlotRange -> {-1.2, 1.2},
         PlotStyle -> {{}, Dashing[{0.03}], GrayLevel[1]},
         AspectRatio -> Automatic, Axes -> None],
         {x, 2, 5, 0.2}];
```

Before showing some of the output of this command, a few words of explanation are in order. As in our earlier example, the curves of intersection are naturally given by implicit equations. So, we used the `ImplicitPlot` command. We drew the intersection of the plane $x =$ constant with the torus in the standard plot style, and the intersection of the plane with the cylinder in a dashed style. We also want to make sure that each graph has the same width and height. Getting the height right is easy: we simply set `PlotRange` to a value that will show all the detail we want. Next, we set `AspectRatio` to `Automatic` to ensure that *Mathematica* uses the same scale on both axes. Unfortunately, that's not quite enough to control the width of the graphs. Since we are using `ImplicitPlot`, *Mathematica* will trim the horizontal axis to fit exactly the set of values that it finds on the implicitly defined curve. We've gotten around that restriction by plotting an extra invisible curve: a circle of radius 5 centered at the origin. (It's invisible because we set a `GrayLevel` of 1, which corresponds to drawing a white curve on a white background.)

We will not show all the output of this command. Since x runs from 2 to 5 by increments of 0.2, it actually produces graphs of 16 distinct cross sections. (If you run this command in a Notebook, then you can instruct *Mathematica* to play the sequence of graphs as a movie: Click on the cell bracket that encloses the entire set of graphs, and then select the menu item `Cell:Animate Selected`

`Graphics.`) We have extracted six representative samples to show you:

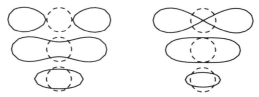

As we hoped, we can use these pictures of the cross sections to make a decision: we do **not** want to save the x variable for last when we integrate. The shapes of the intersections are vastly different for different values of x. Such a variation in cross sectional shapes makes it extremely difficult to describe the region simply. We should instead look at the cross sections we get by holding one of the other variables constant. For example, if we slice through the solid with the planes corresponding to $y = $ constant, then all the cross sections look roughly like this:

```
In[37]:= ImplicitPlot[Evaluate[{toreqn, cyleqn} /. y->0.5],
        {x, -5, 5}, {z, -1, 1}, PlotRange -> {-1.2, 1.2},
        PlotStyle -> {{}, Dashing[[0.03]]], Axes -> None,
        AspectRatio -> Automatic];
```

In this graph, y is constant. The horizontal axis is the x-axis; the vertical axis is the z-axis. This suggests that we should integrate in x-z-y order. We can determine the x-values by solving the torus equation for x as a function of y and z.

```
In[38]:= xsol = Solve[toreqn, x]
```
$$Out[38]= \left\{\left\{x \to -\sqrt{17 - y^2 - z^2 - 8\sqrt{1 - z^2}}\right\},\right.$$
$$\left\{x \to \sqrt{17 - y^2 - z^2 - 8\sqrt{1 - z^2}}\right\},$$
$$\left\{x \to -\sqrt{17 - y^2 - z^2 + 8\sqrt{1 - z^2}}\right\},$$
$$\left.\left\{x \to \sqrt{17 - y^2 - z^2 + 8\sqrt{1 - z^2}}\right\}\right\}$$

The two negative solutions correspond to the intersections to the left of the origin; the other two correspond to the intersections to the right of the origin. By symmetry, we only need the piece on the right.

```
In[39]:= xlo = x /. xsol[[2]];
        xhi = x /. xsol[[4]];
```

We can determine the limits of integration for z by observing that the cylinder intersects the y-z plane in a circle of radius 1. So, the following command produces a graph of the piece of the solid intersection to the right of the origin.

```
In[40]:= viewSolid[{y, -1, 1},
         {z, -Sqrt[1-y^2], Sqrt[1-y^2]}
         {x, xlo, xhi}];
```

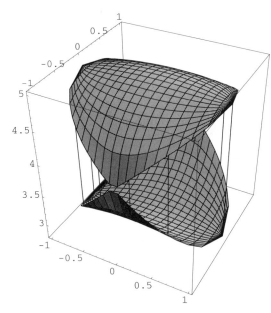

Finally, the volume of the total intersection is given by the integral:

$$2 \int_{-1}^{1} \int_{-\sqrt{1-y^2}}^{\sqrt{1-y^2}} \int_{xlo}^{xhi} 1 \, dx \, dy \, dz.$$

Mathematica cannot evaluate this integral symbolically, so we will compute it numerically. Since numerical methods in three dimensions can take a long time, we will compute the integral in two steps. First, we can compute the integral with respect to x symbolically. We then switch to numerical methods for the remaining integrations.

```
In[41]:= intermediate = 2*Integrate[1, {x, xlo, xhi}];
         total = NIntegrate[intermediate, {y, -1, 1},
            {z, -Sqrt[1-y^2], Sqrt[1-y^2]}]
```

```
Out[41]= 10.7633
```

EXERCISE. Estimate the volume using the picture generated by In[40] and explain why this answer is reasonable.

Cylindrical Coordinates

The problem we have just solved (computing the volume of the intersection of a solid torus with a solid cylinder) can also be done, perhaps more naturally, in cylindrical coordinates. Without giving full details, we briefly indicate how to do this. To begin, we need the equations of the torus and cylinder in cylindrical coordinates. The equation of the torus came to us in the form $(r - 4)^2 + z^2 = 1$, but we need to convert the equation of the cylinder.

```
In[42]:= newtoreqn = ((r-4)^2 + z^2 == 1);
```

```
In[43]:= newcyleqn = cyleqn /. {y -> r*Sin[theta]}
```

Out[43]= $z^2 + r^2 \operatorname{Sin[theta]}^2 == 1$

Note that the equation of the torus imposes no constraints at all on the angular variable θ, while the equation of the cylinder tells us that in the region of integration, $|r \sin \theta| \leq 1$, or $|\sin \theta| \leq (1/r)$. We also know from the equation of the torus that $|r - 4| \leq 1$, or that $3 \leq r \leq 5$. By symmetry, the volume we want is 8 times the volume of the portion of the region of intersection in the first octant. In this portion of the region, for fixed r, θ ranges from 0 to $\arcsin(1/r)$. Then for fixed r and θ, z ranges from 0 to the smaller of the two z-values given by the equations of the torus and the cylinder. Recall that in cylindrical coordinates, we need to include a factor of r in the integrand. So, the desired volume is

```
In[44]:= 8*NIntegrate[r, {r, 3, 5}, {theta, 0, ArcSin[1/r]},
            {z, 0, Sqrt[Min[1-(r-4)^2, 1-r^2 Sin[theta]^2]]}]
```

This results in lots of warning messages about slow convergence, but eventually produces the output:

Out[44]= $10.7633 + 3.7715 \times 10^{-29} \mathrm{I}$

the same as what we obtained before, except for round-off error.

More General Changes of Coordinates

Some multiple integrals can be computed more easily by changing from the standard rectangular coordinates to another coordinate system. The best-known other systems are polar, cylindrical, and spherical coordinates. But in this section, we explain how to use *Mathematica* to set up and evaluate integrals involving a more general change of coordinates.

For simplicity, we will confine our discussion to two dimensions. We may want to change variables in the two-dimensional integral

$$\int\int_R h(x, y) \, dx \, dy$$

for one of two reasons: either to simplify the integrand, or to simplify the region. For example, let's try to compute the area of the region R in the first quadrant enclosed by the two hyperbolas

$$x^2 - y^2 = 1, \qquad x^2 - y^2 = 4,$$

and the two parabolas

$$y = 9 - x^2, \qquad y = 16 - x^2.$$

Here is a picture of the region.

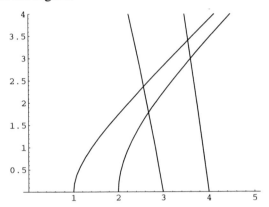

It is clear that solving the problem in rectangular coordinates—either vertically or horizontally—will require dividing the region into three separate pieces. There is an alternative, however. The bounding curves in this example come in pairs. Each pair consists of two level curves of some function of two variables. We can make a change of variables using those auxiliary functions:

$$u = x^2 - y^2,$$
$$v = x^2 + y.$$

When we change variables like this, the formula for the area of the region becomes

$$\text{area}(R) = \iint_R dx \, dy = \iint_S |J(u, v)| \, du \, dv,$$

where S is the region in the u-v plane that R transforms into under the change of variables, and $J(u, v)$ is the reciprocal of the determinant of the Jacobian matrix

$$\begin{pmatrix} \partial u / \partial x & \partial u / \partial y \\ \partial v / \partial x & \partial v / \partial y \end{pmatrix}.$$

Three steps are required to compute the integral with this change of coordinates. First, we need to describe the region S. In our example, we chose coordinates to make that task particularly easy: S is the rectangle $1 \le u \le 4$, $9 \le v \le 16$. Next, we need to compute the determinant of the Jacobian matrix.

```
In[45]:= f = x^2 - y^2;
         g = x^2 + y;
```

```
In[46]:= jake = Det[{{D[f, x], D[f, y]}, {D[g, x], D[g, y]}}]
```

Out[46]= $2x + 4xy$

Finally, we need to rewrite the inverse of the Jacobian in terms of the variables u and v. So, we need to solve the change of coordinate formulas for x and y.

```
In[47]:= backsols = Solve[{u == f, v == g}, {x, y}]
```

$$\text{Out[47]}= \left\{\left\{y \to \frac{1}{2}(-1 - \sqrt{1 - 4u + 4v}),\ x \to -\frac{\sqrt{1 + 2v + \sqrt{1 - 4u + 4v}}}{\sqrt{2}}\right\},\right.$$

$$\left\{y \to \frac{1}{2}(-1 - \sqrt{1 - 4u + 4v}),\ x \to \frac{\sqrt{1 + 2v + \sqrt{1 - 4u + 4v}}}{\sqrt{2}}\right\},$$

$$\left\{y \to \frac{1}{2}(-1 + \sqrt{1 - 4u + 4v}),\ x \to -\frac{\sqrt{1 + 2v - \sqrt{1 - 4u + 4v}}}{\sqrt{2}}\right\},$$

$$\left.\left\{y \to \frac{1}{2}(-1 + \sqrt{1 - 4u + 4v}),\ x \to \frac{\sqrt{1 + 2v - \sqrt{1 - 4u + 4v}}}{\sqrt{2}}\right\}\right\}$$

There are four solutions, since the hyperbolas and parabolas actually meet once in each of the four quadrants. The solution in the first quadrant is the last of these.

```
In[48]:= jacobianuv = (1/jake) /. Last[backsols];
```

Mathematica cannot compute the integral symbolically, but it can still get a numerical answer.

```
In[49]:= NIntegrate[jacobianuv, {u, 1, 4}, {v, 9, 16}]
```

Out[49]= 0.544171

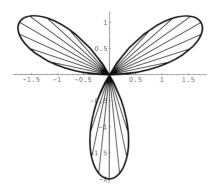

Problem Set H

MULTIPLE INTEGRALS

In these problems, when computing the value of a definite integral, you should first try to use **Integrate**. If this fails, then use **NIntegrate**. If **Integrate** succeeds and gives a symbolic answer, use **N** to obtain a numerical value. You may use the programs from Chapter 8, found on the accompanying disk.

1. Each of the following problems contains a double integral. In each case, you should first use the **verticalRegion** program from Chapter 8 to plot the (shaded) region over which the integral extends. Then compute the value of the integral.

    (a) $\displaystyle\int_1^2 \int_0^x y\sqrt{x^2 - y^2}\, dy\, dx$.

    (b) $\displaystyle\int_0^1 \int_{x^2}^x xy\, dy\, dx$.

    (c) $\displaystyle\int_1^e \int_{\ln(x)/x}^{1/x} e^{xy}\, dy\, dx$.

    (d) $\displaystyle\int_{1.6}^{4.7} \int_{\tan x}^{e^x} \cos e^x\, dy\, dx$.

2. Each of the following problems contains a double integral. In each case, you should first convert to polar coordinates. Next, use the **polarRegion** program

from Chapter 8 to plot the (shaded) region over which the integral extends. Finally, compute the value of the integral.

(a) $\iint_R x^2 y^2 \, dA$, where R is the inside of the cardioid $r = 2(1 + \sin \theta)$.

(b) $\iint_R \cos(x^2 + y^2) \, dA$, where R is defined by $0 \leq \theta \leq \pi/2$ and $0 \leq r \leq \sin^2 \theta$.

(c) $\int_{3/\sqrt{2}}^{3} \int_{0}^{\sqrt{9-x^2}} (x^2 + y^2) \, dy \, dx$.

(d) $\int_{0}^{1} \int_{\sqrt{x-x^2}}^{\sqrt{1-x^2}} (x^2 + y^2)^{3/2} \, dy \, dx$.

3. A horizontally simple region R is bounded below by the horizontal line $y = c$, above by the horizontal line $y = d$, on the left by the curve $x = f(y)$, and on the right by the curve $x = g(y)$. It is natural to integrate over horizontally simple regions with respect to x first, and then y.

(a) Write a program called **horizontalRegion** that takes c, d, f, and g as its arguments. The output of this program should be a graph that shades the region R with horizontal lines. (Hint: Modify the **verticalRegion** program from the chapter to use **ParametricPlot** instead of **Plot**, where you view the curve $x = f(y)$ as a parametric curve $(f(y), y)$ parametrized by y.)

(b) Use the **horizontalRegion** program to shade the regions over which each of the following integrals is being computed; then evaluate the integrals:

(i) $\int_{0}^{2} \int_{y/2}^{y} e^{x^2} \, dx \, dy$.

(ii) $\int_{0}^{1} \int_{-\sqrt{1-y^2}}^{\sqrt{1-y^2}} yx^4 \, dx \, dy$.

(iii) $\int_{-2}^{2} \int_{1-\sqrt{2-y}}^{y/2} xe^x \, dx \, dy$.

4. Each of the following double integrals extends over a region that is both vertically simple and horizontally simple. In each case, you should:

(i) Plot the shaded region corresponding to the integral as given.
(ii) Compute the value of the given integral.
(iii) Reverse the order of integration. Plot the new shaded region.
(iv) Compute the value of the integral with the order of integration reversed.

(Note: In some cases, *Mathematica* will not graph the entire region unless you redisplay the graph using **Show** with **PlotRange -> All**.)

(a) $\displaystyle\int_0^1 \int_{\sqrt[3]{y}}^1 \sqrt{1+x^4}\, dx\, dy.$

(b) $\displaystyle\int_0^{\pi/2} \int_0^{\cos x} x^2\, dy\, dx.$

(c) $\displaystyle\int_{-5}^4 \int_{2-\sqrt{4-y}}^{(y+2)/3} e^{x^2-4x+y}\, dx\, dy.$

5. Each part of this problem describes a region in the plane. Plot the indicated region. Then find the area. You may need to use `Solve` or `FindRoot` to find the points of intersection of the curves.

   (a) The region bounded by the parabolas $y = 3x^2$ and $y = 5x - 2x^2$.

   (b) The region bounded by the ellipse $4x^2 + y^2 = 4$.

   (c) The total region bounded by the curves $y = \cos x$ and $y = 4x^2 - x^4$.

   (d) The total region bounded by the curve $y = e^{-x^2}\cos 10x$ and the line $y = 1 - x$.

6. Each part of this problem describes a region in the plane. Plot (and shade) the indicated region. Then find the area. You may need to use `Solve` or `FindRoot` to find the points of intersection of the curves.

   (a) The region inside the rose $r = 2\sin 3\theta$.

   (b) The region inside the lemniscate $r^2 = 2\cos 2\theta$ and outside the circle $r = 1$.

   (c) The region inside the cardioid $r = 1 + \cos\theta$ and outside the circle $r = 1/2$.

   (d) The region inside the circle of radius 8 and outside the loop of the curve defined implicitly by $8y^2 = x^3 + 8x^2$.

7. Each of the following problems describes a region R and a function $f(x, y)$. In each case, you should graph the shaded region, and then compute the value of the integral

$$\iint_R f(x, y)\, dA.$$

You must decide: (i) whether to use rectangular or polar coordinates; and (ii) in which order to integrate. Your shaded graph should coincide with the chosen method for evaluating the integral.

   (a) $f(x, y) = e^{7y-x}$ and R is the region bounded by $y^2 = x - 12$ and $y^2 = 30 - x$.

   (b) $f(x, y) = 1/\sqrt{x^2 + y^2}$ and R is the region to the right of $x = 0$ which is inside the ellipse $x^2 + 4y^2 = 9$ and outside the circle $(x - 1)^2 + y^2 = 1$.

(c) $f(x, y) = x^2 + y^2$ and R is the bounded region between the graphs of $y = \sin x$ and $3y + x = 3$.

8. Each of these problems presents a triple integral. Use the `viewSolid` program from Chapter 8 to sketch the region over which the integral extends, and then evaluate the integral.

(a) $\displaystyle \int_0^1 \int_{x^3}^{x^2} \int_0^{(x+y)/4} z\sqrt{x}\, dz\, dy\, dx.$

(b) $\displaystyle \int_0^1 \int_0^{\sqrt{1-y^2}} \int_{x^2+y^2-1}^{1-x^2-y^2} e^y \sqrt{1-y^2}\, dz\, dx\, dy.$

(c) $\displaystyle \int_0^{\pi/6} \int_0^z \int_0^{y+z} (1 + y^2 z^3 \sin xz)\, dx\, dy\, dz.$

(d) $\displaystyle \int_0^1 \int_0^{x^2} \int_0^{x^3-y^3} x^2 y^2 e^z\, dz\, dy\, dx.$

9. Each of these problems presents a triple integral. Use the `viewSolid` program from Chapter 8 to sketch the region D over which the integral extends, and then evaluate the integral. You should be able to determine the limits of integration directly from the description of D. In (c) and (d), it may help to think about planes that contain three of the given points.

(a) $\iiint_D xyz\, dV$, where D is the region in the first octant bounded above by the hemisphere $z = \sqrt{4 - x^2 - y^2}$ and on the sides and the bottom by the coordinate planes.

(b) $\iiint_D xz\, dV$, where D is the region in the first octant bounded above by the sphere $x^2 + y^2 + z^2 = 9$, below by the plane $z = 0$, and on the sides by the planes $x = 0$, $y = 0$, and the cylinder $x^2 + y^2 = 4$.

(c) $\iiint_D yz^2 e^{xy}\, dV$, where D is the prism with vertices $(0, 0, 1)$, $(0, 1, 2)$, $(0, 1, 0)$, $(1, 0, 1)$, $(1, 1, 2)$, and $(1, 1, 0)$.

(d) $\iiint_D \log(z)/z\, dV$, where D is the tetrahedron with vertices at $(1, 0, 2)$, $(1, 1, 2)$, $(1, 1, 1)$, and $(2, 0, 2)$.

10. Each part of this problem describes a solid region. Graph the region. Based on the graph, estimate the volume. (Don't just write down a number; give a brief plausibility argument.) Then compute an integral to find the actual volume.

(a) The solid region between the saddle $z = xy/2$ and the cone $z^2 = x^2 + 2y^2$ that lies above the portion in the first quadrant of the disk whose boundary is $x^2 + y^2 = 1$.

(b) The solid region bounded above by the graph of $z = y^2/4$, and whose base in the x-y plane is the bounded region between the curves $x = \sin y$ and $x = 3y - y^2$.

(c) A right pyramid with height h and a square base having sides of length s. (Hint: Place the base of the pyramid in the x-y plane, with center at the origin and vertices on the axes. For purposes of plotting a sample, pick specific values of s and h. Using symmetry, just compute the volume of the part in the first octant.)

(d) The region bounded above by the graph of the function $z = \frac{1}{2} + x^2 - y^2$, on the sides by the cylinder $x^2 + y^2 = x$, and below by the x-y plane.

11. This problem is concerned with regions defined in cylindrical coordinates.

(a) Modify the `viewSolid` program from Chapter 8 to construct a program called `viewCylindrical`. The arguments to `viewCylindrical` should be (except for the integrand) the same as those needed to compute the triple integral

$$\int_\alpha^\beta \int_{f(\theta)}^{g(\theta)} \int_{F(r,\theta)}^{G(r,\theta)} h(r, \theta, z) r \, dz \, dr \, d\theta.$$

(b) Use your program to sketch the regions over which each of the following integrals extends. Describe the resulting regions in words.

(i) $\displaystyle\int_0^\pi \int_0^1 \int_0^{\sqrt{1-r^2}} r \, dz \, dr \, d\theta.$

(ii) $\displaystyle\int_0^{2\pi} \int_0^1 \int_0^r r \, dz \, dr \, d\theta.$

(iii) $\displaystyle\int_0^{2\pi} \int_0^3 \int_{1-(r^2/9)}^{\sqrt{r^2+4}} r \, dz \, dr \, d\theta.$

12. Each of these problems presents a triple integral in cylindrical coordinates. Sketch the region over which the integral extends, and then evaluate the integral.

(a) $\displaystyle\int_0^{2\pi} \int_0^1 \int_r^1 zr^3 \cos^2 \theta \, dz \, dr \, d\theta.$

(b) $\displaystyle\int_0^{2\pi} \int_0^1 \int_{-\sqrt{4-r^2}}^{\sqrt{4-r^2}} z^2 r \, dz \, dr \, d\theta.$

(c) Find the volume of the region under the cone $z = 6 - r$ whose base is bounded by the cardioid $r = 3(1 + \cos \theta)$.

13. This problem is concerned with regions defined in spherical coordinates.

(a) Modify the `polarRegion` program from Chapter 8 to construct a program called `viewSpherical`. The arguments to `viewSpherical` should be (except for the integrand) the same as those needed to compute the triple integral

$$\int_a^b \int_{f(\theta)}^{g(\theta)} \int_{F(\theta,\phi)}^{G(\theta,\phi)} \rho^2 \sin\phi \, d\rho \, d\phi \, d\theta.$$

(Hints: You might want to read about the `SphericalPlot3D` command in the `Graphics'ParametricPlot3D'` package. Also, since labeling conventions differ, you should allow the user to choose the names for the angular coordinates.)

(b) Use your program to sketch the regions over which each of the following integrals extends. Here ρ is the distance to the origin, ϕ is the colatitude, and θ is the angular variable in the x-y plane.

(i) $\displaystyle\int_0^{\pi/2} \int_0^{\pi/2} \int_1^3 \rho^2 \sin\phi \, d\rho \, d\phi \, d\theta.$

(ii) $\displaystyle\int_0^{2\pi} \int_0^{\pi/4} \int_1^{2\sec\phi} \rho^2 \sin\phi \, d\rho \, d\phi \, d\theta.$

(iii) $\displaystyle\int_0^{2\pi} \int_0^{\pi/6} \int_0^{\sqrt{\sec(2\phi)}} \rho^2 \sin\phi \, d\rho \, d\phi \, d\theta.$

14. Each of these problems presents a triple integral in spherical coordinates. Sketch the region over which the integral extends, and then evaluate the integral. Here ρ is the distance to the origin, ϕ is the colatitude, and θ is the angular variable in the x-y plane.

(a) $\displaystyle\iiint_D (x^2 + y^2) \, dV$, where D is the ball $x^2 + y^2 + z^2 \le 1$.

(b) The volume inside the cone $3(x^2 + y^2) = z^2$ and below the upper sheet of the hyperboloid $z^2 - x^2 - y^2 = 1$.

(c) The volume of a "slice of an apple," that is, the smaller of the two regions bounded by the sphere $\rho = 3$ and the half-planes $\theta = 0$ and $\theta = \pi/6$.

(d) The volume of the solid region bounded below by the x-y plane and above by the surface $\rho = 1 + \cos\phi$.

15. You can use multiple integrals to compute surface areas. Suppose Σ is a surface patch in the sense of Chapter 6, *Geometry of Surfaces*, defined by a parametrization $\sigma: D \to \mathbf{R}^3$. Then the area of Σ is:

$$\text{area}(\Sigma) = \iint_D J(u, v) \, du \, dv,$$

where the Jacobian factor $J(u, v)$ measures the local stretching done by σ. The formula for J is derived as follows. As u ranges between u_0 and $u_0 + \Delta u$ and as v

ranges between v_0 and $v_0 + \Delta v$, $\sigma(u, v)$ approximately traces out a parallelogram, with vertices at $\sigma(u_0, v_0)$, $\sigma(u_0, v_0) + \sigma_u(u_0, v_0)\Delta u$, $\sigma(u_0, v_0) + \sigma_v(u_0, v_0)\Delta v$, and at $\sigma(u, v) + \sigma_u(u_0, v_0)\Delta u + \sigma_v(u_0, v_0)\Delta v$. The area formula in Chapter 2 gives the area of this parallelogram as $\|\sigma_u(u_0, v_0)\Delta u \times \sigma_v(u_0, v_0)\Delta v\|$. So, adding the areas and passing to the limit yields the formula:

$$J(u, v) = \|\sigma_u \times \sigma_v\|;$$

in other words, the Jacobian factor equals the length of the normal vector **N** that we studied in Chapter 6.

(a) Show that the formula for $J(\phi, \theta)$ reduces to $a^2 \sin \phi$ in the case of a piece of a sphere of radius a parametrized by spherical coordinates ϕ and θ:

$$\sigma(\phi, \theta) = (a \cos \theta \sin \phi, \, a \sin \theta \sin \phi, \, a \cos \phi).$$

(b) Show that the formula for $J(x, y)$ reduces to $\sqrt{f_x^2 + f_y^2 + 1}$ in the case of a Monge patch parametrization $\sigma(x, y) = (x, y, f(x, y))$ with parameters x and y.

(c) Show that the formula for $J(z, \theta)$ reduces to $r(z)\sqrt{1 + r'(z)^2}$ in the case of a surface of revolution parametrized by cylindrical coordinates z and θ:

$$\sigma(z, \theta) = (r(z) \cos \theta, \, r(z) \sin \theta, \, z).$$

(d) Sketch each of the following surfaces and compute the surface area:

(i) The portion of the paraboloid $z = 9 - x^2 - y^2$ above the x-y plane.

(ii) The portion of the sphere $x^2 + y^2 + z^2 = 4$ that is above the x-y plane and inside the cylinder $x^2 + y^2 = 1$.

(iii) The portion of the sphere $x^2 + y^2 + z^2 = 9$ that is inside the paraboloid $x^2 + y^2 = 8z$.

(iv) The torus whose equation in cylindrical coordinates is $(r - 4)^2 + z^2 = 1$.

16. This problem is concerned with the solid region bounded by the surface

$$x^{2/3} + y^{2/3} + z^{2/3} = 4.$$

Because the solid is symmetric, we will only look at the portion D lying in the first octant.

(a) Graph the projection of D onto the first quadrant of the x-y plane.

(b) Find the area of the projected region.

(c) Graph the solid D. (You may need to `Show` the graph with `PlotRange -> All`.)

(d) From the graph, estimate the volume of D.

(e) Finally, compute the volume of the solid D. *Mathematica* cannot symbolically compute the volume integral that you get from rectangular coordinates. Make a change of variables to find the volume. Then check your answer using numerical integration in rectangular coordinates.

17. In this problem, we consider the intersection of the two cylinders $x^2 + y^2 = 1$ and $x^2 + z^2 = 1$.

(a) Graph the two cylinders on the same axes.

(b) Use the methods of Chapter 8 to draw cross sections of the region bounded by the two cylinders. Based on your results, choose an order of integration and compute the volume of the region.

18. In this problem, we consider the intersection of the cylinder $y^2 + (z - 2)^2 = 1$ with the cone $z^2 = x^2 + y^2$.

(a) Graph both surfaces on the same axes.

(b) Use the methods of Chapter 8 to draw cross sections of the region bounded by the cylinder and the cone. Based on your results, choose an order of integration and compute the volume of the region.

19. In this problem, you are asked to compute some areas that are either difficult or impossible to compute in rectangular coordinates. In each part, first draw a picture of the region being discussed. Then compute the area of the region using integration in an appropriate coordinate system.

(a) The region in the first quadrant of the x-y plane that lies between the hyperbolas $xy = 1$ and $xy = 2$ and between the lines $y = x$ and $y = 4x$.

(b) The region in the first quadrant of the x-y plane that lies between the hyperbolas $x^2 - y^2 = 1$ and $x^2 - y^2 = 2$ and between the circles $x^2 + y^2 = 9$ and $x^2 + y^2 = 16$.

20. Sketch the region in the first quadrant of the x-y plane that lies between the circles $x^2 + y^2 = \frac{1}{9}$ and $x^2 + y^2 = \frac{4}{9}$ and between the two parabolas $y = x^2$ and $y = \sqrt{x}$. Then compute the area of the region using a double integral in an appropriate coordinate system. (Hint: You can use the symmetry of the region in computing the area.)

21. In this problem, we will introduce a numerical algorithm, called the *Monte Carlo method*, that can be used to estimate the values of complicated multiple integrals. The reason for the name is that the algorithm relies on a random process somewhat akin to the spinning of a roulette wheel. To illustrate the

method, we will use it to estimate vol $R = \iiint_R dV$, where R is a possibly complicated region in $\mathbf{R}^3$. The first step is to enclose R in a rectangular box, say

$$B = \{x_0 \le x \le x_1, \, y_0 \le y \le y_1, \, z_0 \le z \le z_1\}.$$

Then the volume of R can be computed as

$$\text{vol } R = p \, \text{vol } B = p(x_1 - x_0)(y_1 - y_0)(z_1 - z_0),$$

where p is the fraction of the volume of B occupied by the region R.

What makes the method work is that the fraction p can be interpreted as the probability that a point chosen "at random" in B will lie in R. So if we choose a very large number n of points "randomly" in B (by picking x to be a random number in the interval $[x_0, x_1]$, y to be a random number in the interval $[y_0, y_1]$, and z to be a random number in the interval $[z_0, z_1]$), the Law of Large Numbers says that the number of points that land in R, divided by the number of trials n, will be a good approximation to p. In fact, one can also show, using the Central Limit Theorem in probability theory, that the error in the approximation should behave like $1/\sqrt{n}$. This means convergence to the true answer is somewhat slow, as numerical methods go, but even when R is complicated, the method at least gives a reasonable estimate of the value of the triple integral. Here is a *Mathematica* program for implementing the algorithm:

```
volMonteCarlo[region_, {x0_, x1_}, {y0_, y1_},
  {z0_, z1_}, n_] := Module[{sum}, sum := 0;
  Do[sum = sum + region[Random[Real, {x0, x1}],
    Random[Real, {y0, y1}],
    Random[Real, {z0, z1}]], {n}];
  sum*(x1 - x0)*(y1 - y0)*(z1 - z0)/n]
```

Here the command `Random[Real, {a, b}]` selects a random number between a and b, and the `Do` command repeats an operation a certain number of times. In this case, we use n iterations, and at each step we add either 1 or 0 to the running sum, depending on whether or not our randomly selected point lies in R. Here is a sample usage of the program. Typing

```
sphere[x_, y_, z_] = If[x^2 + y^2 + z^2 < 1, 1, 0];
volMonteCarlo[sphere, {-1, 1}, {-1, 1},
  {-1, 1}, 10000] // N
```

yields the answer 4.1952.

(a) Run the program a few more times with different values of n to estimate the volume of (the region bounded by) a sphere of radius 1. Compare with the numerical value of the exact answer, $4\pi/3$, and study how the error behaves with n. (Warning: Don't take n TOO large with this program or it will take an unmanageable amount of time. $n = 40000$ is about the upper limit on most computers.)

(b) Run the program several times with different values of n to estimate the volume of the region bounded by the astroidal sphere of radius 1:

$$R = \left\{ x^{2/3} + y^{2/3} + z^{2/3} \leq 1 \right\}.$$

(As you may have noticed in Problem 16, this is the sort of problem where numerical algorithms can be useful, because the integral for the volume of R is quite hard to evaluate, even numerically, without a change of variables.) Just one word of caution: fractional exponents can lead to trouble in *Mathematica*, since they are evaluated as complex numbers. To avoid difficulties, take absolute values and use

```
astroid[x_,y_,z_] = If[Abs[x^(2/3)] + Abs[y^(2/3)] +
    Abs[z^(2/3)] < 1, 1, 0]
```

Check your answer for the volume against the results of an exact calculation with the change of variables $x = u^3$, $y = v^3$, $z = w^3$.

(c) By changing the definition of `region`, you can modify the same program to compute more complicated triple integrals of the form $\iiint_R f(x, y, z)\, dV$. Use the program to estimate the mass of a sphere of radius 1 with mass density $\delta(x, y, z) = x^2 y^2 z^2$. Check your answer for the mass against the results of an exact calculation in spherical coordinates. (If you do this right you will see a curious connection with part (b).)

(d) Use the Monte Carlo method to estimate the volume of the region inside the sphere $x^2 + y^2 + z^2 = 4$ and between the two sheets of the hyperboloid $z^2 - (x^2 + y^2) = 1$.

Glossary of Some Useful *Mathematica* Objects

Commands

`CylindricalPlot3D` Plots a surface given in cylindrical coordinates.

`Do` Repeats a command a certain number of times.

`Integrate` Computes single or multiple integrals.

`Line` Creates a line as a graphics object.

`Module` Organizes a number of commands or expressions into a single unit, with specified local variables.

`NIntegrate` Computes single or multiple integrals numerically.

`PolarPlot` Plots a curve given in polar coordinates.

`Random` Selects a random number.

`SphericalPlot3D` Plots a surface given in spherical coordinates.

Options and Directives

`Method` Option for `NIntegrate`, specifying which algorithm to use.

`PlotRange` When set to `All`, displays all of a plot.

`Thickness` Specifies (relative) thickness of curves and lines.

Packages

`Calculus`VectorAnalysis`` Contains commands (such as `JacobianDeterminant`) useful for changing coordinates, especially in multiple integrals.

`Graphics`Graphics`` Needed for `PolarPlot`.

`Graphics`ParametricPlot3D`` Needed for `CylindricalPlot3D` and for `SphericalPlot3D`.

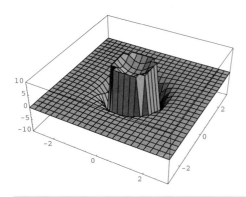

Chapter 9

PHYSICAL APPLICATIONS OF VECTOR CALCULUS

This chapter represents the culmination of multivariable calculus. We investigate the remarkable physical applications of vector calculus that provided the original motivation for the development of this subject in the seventeenth, eighteenth, and nineteenth centuries. The vector fields that we examine arise naturally in celestial mechanics, electromagnetism, and fluid flow. We will use the basic concepts of vector calculus to derive fundamental laws of physics in these subjects. In the attached Problem Set I, you will have a chance to use *Mathematica* to solve some interesting physical problems that would be difficult or impossible to tackle with pencil and paper alone.

Motion in a Central Force Field

In Chapter 4, *Kinematics*, we studied the motion of an object in a central force field as an application of the theory of curves in 3-space. We now reexamine the same subject from the point of view of the calculus of vector fields. We begin with the study of the motion of a *particle*, that is, of an object whose size is negligible compared to the distances over which it moves. For all practical purposes we may assume the object is located at a single point in space. In fact, the theory we develop does not apply very well to the objects usually known as particles in physics, such as electrons, photons, and positrons. True particles are so small that classical physics doesn't provide a very good approximation to their actual behavior. They need to be studied with quantum mechanics, which involves a more complicated vector calculus than we will study here. The theory we develop, however, works quite well when applied to "celestial particles" such as planets and comets moving around a star.

We assume that the location of our particle is represented by a *position vector* $\mathbf{r} = x\mathbf{i} + y\mathbf{j} + z\mathbf{k}$ in $\mathbf{R}^3$, which is a function of the time t. We also assume the particle obeys *Newton's law of motion*, which says that $\mathbf{F} = m\mathbf{a} = m\ddot{\mathbf{r}}$, where $\mathbf{F}$ is the force on the particle, m is its mass, $\mathbf{a}$ is its acceleration, and we use dots to indicate time derivatives.

Suppose that the force on the particle is always a scalar multiple of the position vector, with the scalar factor a function of $r = \|\mathbf{r}\|$, but not of time. This amounts to assuming three things:

- the magnitude of the force on the particle only depends on the distance $r = \|\mathbf{r}\|$ from the origin;
- the force on the particle always points either toward or away from the origin; and
- whether the force points inward or outward (i.e., is *attractive* or *repulsive*) also depends only on $r = \|\mathbf{r}\|$.

When these conditions are satisfied, we say the object is moving in a *central force field*. In symbols, $\mathbf{F} = f(r)\mathbf{r}$. The classical example occurs when $\mathbf{r}(t)$ represents the position of a planet in a solar system with the sun at the origin. We assume the mass m of the planet is negligible compared to the mass M of the sun, so that the sun remains fixed. The planet orbits the sun under the influence of gravity, which is an attractive force always pointing inward toward the sun. *Newton's law of gravitation* says that

$$\mathbf{F} = -\frac{GMm}{\|\mathbf{r}\|^3}\mathbf{r}, \tag{1}$$

where G is a universal constant. The minus sign occurs because gravity is attractive, hence tends to pull objects in the direction of decreasing r. As we pointed out in Chapter 4, Newton's law of gravitation (1) is an *inverse square law*, since

$$\|\mathbf{F}\| = \frac{GMm}{\|\mathbf{r}\|^2}.$$

For the time being, let's see how far we can progress without assuming (1), since there are examples of central force fields in physics where the dependence of $\mathbf{F}$ on $\mathbf{r}$ is different. We assume only that $\mathbf{F} = f(r)\mathbf{r}$. Applying Newton's law of motion, we arrive at the fundamental differential equation governing the motion of the object:

$$\mathbf{F} = f(r)\mathbf{r} = m\ddot{\mathbf{r}}, \tag{2}$$

or

$$\dot{\mathbf{v}} = \ddot{\mathbf{r}} = \frac{f(r)}{m}\mathbf{r} = g(r)\mathbf{r}.$$

Let's take the curl of the vector field $\mathbf{F}(\mathbf{r}) = f(r)\mathbf{r} = f(r)(x\mathbf{i} + y\mathbf{j} + z\mathbf{k})$ that gives the force on the particle. We can do this in *Mathematica* as follows:

```
In[1]:= <<Calculus'VectorAnalysis'
```

```
In[2]:= SetCoordinates[Cartesian[x,y,z]]
```

Out[2]= $\text{Cartesian}[x, y, z]$

```
In[3]:= rvec = {x,y,z}; r = Sqrt[rvec.rvec]
```

Out[3]= $\sqrt{x^2 + y^2 + z^2}$

```
In[4]:= Curl[f[r]*rvec]
```

Out[4]= $\{0, 0, 0\}$

Thus

$$\operatorname{curl}\mathbf{F} = \nabla \times \mathbf{F} = 0. \tag{3}$$

This suggests that $\mathbf{F}$ should be a *conservative* force field, i.e., that $\mathbf{F} = -\nabla\phi$ for some function ϕ. (Vanishing of the curl doesn't quite imply that $\mathbf{F}$ is the gradient of a function, because of possible singularities of $\mathbf{F}$ at the origin.) Indeed, suppose $\phi(\mathbf{r})$ depends only on $r = \|\mathbf{r}\|$, so that we can write $\phi(r)$ without fear of confusion. We compute the gradient:

```
In[5]:= Grad[phi[r]]
```

Out[5]= $\left\{\dfrac{x\,\text{phi}'[\sqrt{x^2 + y^2 + z^2}]}{\sqrt{x^2 + y^2 + z^2}}, \dfrac{y\,\text{phi}'[\sqrt{x^2 + y^2 + z^2}]}{\sqrt{x^2 + y^2 + z^2}}, \right.$

$\left. \dfrac{z\,\text{phi}'[\sqrt{x^2 + y^2 + z^2}]}{\sqrt{x^2 + y^2 + z^2}}\right\}$

The output is equal to $(\phi'(r)/r)\mathbf{r}$, so $\mathbf{F} = -\nabla\phi$ provided we choose ϕ such that

$$\frac{\phi'(r)}{r} = -f(r).$$

In other words, we should choose $\phi(r)$ to be an antiderivative of $-rf(r)$. Such a choice of ϕ is called a *potential function* for the central force field. If $f(r) \to 0$ rapidly enough as $r \to \infty$ so that the following improper integral converges, then we can normalize ϕ by taking

$$\phi(r) = \int_r^\infty x f(x)\,dx. \tag{4}$$

In this case, $\phi(r) \to 0$ as $r \to \infty$.

Finally, define

$$E = \frac{m}{2}\|\mathbf{v}\|^2 + \phi(r), \tag{5}$$

which we think of as a function of t by evaluating it along the path of the particle. This expression is called the *energy* of the particle. The first term is the *kinetic energy*; the second, the *potential energy*. If we express the kinetic energy as a dot product, $\frac{1}{2}m\mathbf{v} \cdot \mathbf{v}$, its time derivative becomes

$$\frac{d}{dt}\left(\frac{m}{2}\mathbf{v} \cdot \mathbf{v}\right) = \frac{m}{2}(\dot{\mathbf{v}} \cdot \mathbf{v} + \mathbf{v} \cdot \dot{\mathbf{v}}) = m\mathbf{a} \cdot \mathbf{v} = \mathbf{F} \cdot \mathbf{v} = f(r)\mathbf{r} \cdot \dot{\mathbf{r}}.$$

Now ϕ was chosen so that $\mathbf{F} = -\nabla\phi$. By the chain rule, the time derivative of the potential energy is:

$$\dot{\phi} = \frac{\partial\phi}{\partial x}\frac{dx}{dt} + \frac{\partial\phi}{\partial y}\frac{dy}{dt} + \frac{\partial\phi}{\partial z}\frac{dz}{dt} = \nabla\phi \cdot \dot{\mathbf{r}} = -\mathbf{F} \cdot \mathbf{v}.$$

This precisely cancels the time derivative of the kinetic energy, and so $\dot{E} = 0$. This is the *law of conservation of energy*; it says that the quantity E remains fixed once and for all.

To summarize, assuming we have an object moving in a central force field according to Newton's laws of motion, we have used vector calculus to derive the law of conservation of energy. We have also demonstrated that the force field must be conservative. Recall that under the same hypotheses, we derived in Chapter 4 the law of conservation of angular momentum, and we showed that the motion of such an object is planar.

Newtonian Gravitation

In the previous section, we discussed Newton's law of gravitation, but only in the special case of a single object moving in a central force field. A more general and interesting problem is the *n-body problem*, the study of the motion of n bodies (which we again assume are "particles," i.e., of negligible size compared to the distances between them), moving under the influence of their mutual gravitational attractive forces. If $\mathbf{r}_i$ denotes the position vector of the ith particle and m_i is its mass ($i = 1, 2, \ldots, n$), then each particle exerts a force on the others given by the inverse square law (1) as discussed above, so there are equations of motion analogous to (2) above. In particular, the force on the ith particle is given by

$$m_i\ddot{\mathbf{r}}_i = -\sum_{j \neq i} Gm_im_j\frac{\mathbf{r}_i - \mathbf{r}_j}{\|\mathbf{r}_i - \mathbf{r}_j\|^3}.$$

While these equations appear quite complicated, a few important features deserve comment. The first feature, which is quite special to gravitation, is the fact that we can cancel m_i from both sides of the equation. (There is an m_i on the right, since the force between the ith and jth particle is proportional to the product of the masses $m_i m_j$; while there is an m_i on the left, since Newton's law of motion says that the force on the ith particle is equal to $m_i \ddot{\mathbf{r}}_i$.) We obtain

$$\ddot{\mathbf{r}}_i = -\sum_{j \neq i} G m_j \frac{\mathbf{r}_i - \mathbf{r}_j}{\|\mathbf{r}_i - \mathbf{r}_j\|^3}, \tag{6}$$

whose right-hand side *no longer involves the mass of the ith particle*. It only involves the position $\mathbf{r}_i$ of the ith particle and the positions and masses of the other particles. We may think of it as a *vector field* (i.e., a vector-valued function of $\mathbf{r}$), evaluated at $\mathbf{r} = \mathbf{r}_i$.

We can use the package `Graphics'PlotField3D'` to visualize this vector field for various configurations of masses. For example, with an object of mass 2 located at $(1, 0, 0)$ and with an object of mass 1 located at $(-1, 0, 0)$, the vector field looks like this:

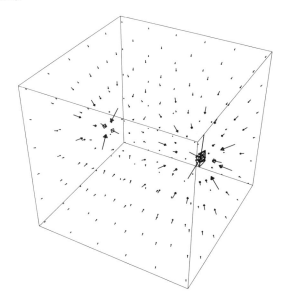

We will now find that we can reformulate the laws of gravitation by considering the flux of the gravitational vector field. Recall that the *flux* of a vector field $\mathbf{V}$ over an oriented surface S is the surface integral $\iint_S \mathbf{V} \cdot \mathbf{n} \, dS$, where $\mathbf{n}$ is the unit "outward pointing" normal vector. Let's take S to be any simple closed piecewise smooth surface (for instance, a sphere, or the surface of a cube) enclosing all of the particles except for the ith. Then the flux through S of the vector field specified by the right-hand side of (6) is given by

$$\iint_S \left(-\sum_{j \neq i} G m_j \frac{\mathbf{r} - \mathbf{r}_j}{\|\mathbf{r} - \mathbf{r}_j\|^3} \right) \cdot \mathbf{n} \, dS = -\sum_{j \neq i} G m_j \iint_S \frac{\mathbf{r} - \mathbf{r}_j}{\|\mathbf{r} - \mathbf{r}_j\|^3} \cdot \mathbf{n} \, dS. \tag{7}$$

To evaluate (7), we only need to compute

$$\iint_S \frac{\mathbf{r} - \mathbf{r}_j}{\|\mathbf{r} - \mathbf{r}_j\|^3} \cdot \mathbf{n} \, dS. \tag{8}$$

If S were a sphere of radius r_0 centered at $\mathbf{r}_j$, then (8) would become

$$
\begin{aligned}
\iint_{\|\mathbf{r} - \mathbf{r}_j\| = r_0} \frac{\mathbf{r} - \mathbf{r}_j}{\|\mathbf{r} - \mathbf{r}_j\|^3} \cdot \mathbf{n} \, dS &= \iint_{\|\mathbf{r}\| = r_0} \frac{\mathbf{r}}{\|\mathbf{r}\|^3} \cdot \mathbf{n} \, dS \\
&= \iint_{\|\mathbf{r}\| = r_0} \frac{\mathbf{n}}{r_0^2} \cdot \mathbf{n} \, dS \\
&= \frac{1}{r_0^2} \iint_{\|\mathbf{r}\| = r_0} dS \\
&= \frac{1}{r_0^2} 4\pi r_0^2 \\
&= 4\pi. \tag{9}
\end{aligned}
$$

Away from the origin, we can compute the divergence of $(1/r^3)\mathbf{r}$ by

```
In[6]:=  Div[(1/r^3)*rvec] // Simplify

Out[6]=  0
```

Thus, away from the point $\mathbf{r} = \mathbf{r}_j$ where the integrand in (8) blows up, we have

$$\operatorname{div} \frac{\mathbf{r} - \mathbf{r}_j}{\|\mathbf{r} - \mathbf{r}_j\|^3} = 0. \tag{10}$$

This means the flux integral (8) is independent of the choice of the surface S. To see that, apply the Divergence Theorem to the shell region between S_1 and S_2, where S_1 and S_2 are simple closed piecewise smooth surfaces around $\mathbf{r}_j$, with S_1 inside S_2. Since (10) holds inside the shell region, the Divergence Theorem gives

$$\iint_{S_2} - \iint_{S_1} = \iint_{S_2 - S_1} = 0,$$

(where for convenience we have suppressed the integrand). Hence (8) is also independent of S, provided S encloses all the points $\mathbf{r}_j$. Putting everything together, we get *Gauss' law*:

$$
\begin{aligned}
\iint_S \left(-\sum_{j \neq i} Gm_j \frac{\mathbf{r} - \mathbf{r}_j}{\|\mathbf{r} - \mathbf{r}_j\|^3} \right) \cdot \mathbf{n} \, dS &= -\sum_{j \neq i} Gm_j 4\pi \tag{11} \\
&= -4\pi G(\text{mass enclosed by } S).
\end{aligned}
$$

In addition, the calculation in (3) shows that the curl of the right-hand side of (6) vanishes (except at the points $\mathbf{r}_j$ where it is undefined).

By extension, we see that according to the laws of Newtonian gravitation, any distribution of masses in space defines a *gravitational field* $\mathbf{G}(\mathbf{r})$, a vector field with the property that the acceleration due to gravity of a particle of arbitrary mass m with position vector $\mathbf{r}$ is given by the vector field $\mathbf{G}(\mathbf{r})$, regardless of the value of m. This vector field has vanishing curl:

$$\mathbf{curl}\ \mathbf{G} = \mathbf{0}, \tag{12}$$

and satisfies Gauss' law:

$$\iint_S \mathbf{G} \cdot \mathbf{n}\, dS = -4\pi G(\text{mass enclosed by } S). \tag{13}$$

Thus vector calculus gives us a very elegant way to reformulate the basic principles of Newtonian gravitation.

These laws apply (as we can see by an appropriate limiting process) also when the distribution of masses in space is continuous rather than discrete. Then we no longer have singularities in $\mathbf{G}(\mathbf{r})$ at point masses. The formulation (13) is particularly useful when the mass distribution is spherically symmetric about the origin, i.e., only depends on $r = \|\mathbf{r}\|$. Then the symmetry implies that $\mathbf{G}(\mathbf{r})$ is a scalar function $g(r)$ of r multiplied by the unit vector $\mathbf{n} = \mathbf{r}/r$ in the direction of $\mathbf{r}$. From (13), with S the sphere of radius r centered at the origin, we have

$$
\begin{aligned}
-4\pi G(\text{mass enclosed by } S) &= \iint_S \mathbf{G} \cdot \mathbf{n}\, dS \\
&= \iint_S g(r)\mathbf{n} \cdot \mathbf{n}\, dS \\
&= g(r) \cdot \text{surface area}(S) \\
&= 4\pi r^2 g(r),
\end{aligned}
$$

and thus

$$\mathbf{G}(\mathbf{r}) = -\frac{G\mathbf{n}}{r^2}(\text{mass enclosed by sphere of radius } r). \tag{14}$$

It is also possible to give an equivalent formulation of Gauss' Law (13) using the Divergence Theorem. If $\rho(\mathbf{r})$ is the density of matter at $\mathbf{r}$, and $\mathcal{R}$ is the solid region bounded by S, then the mass enclosed by S is

$$\iiint_{\mathcal{R}} \rho(\mathbf{r})\, dV.$$

Now (13) can be rewritten using the Divergence Theorem as

$$\iiint_{\mathcal{R}} \text{div}\, \mathbf{G}(\mathbf{r})\, dV = -4\pi G \iiint_{\mathcal{R}} \rho(\mathbf{r})\, dV.$$

This being true for arbitrary regions $\mathcal{R}$, we must have

$$\text{div}\, \mathbf{G} = -4\pi G\rho. \tag{15}$$

Formulas (15) and (13) give a convenient formulation for studying the gravitational field inside bodies which are no longer considered to be particles (see some applications in the exercises). The formulation is also useful for comparison with Maxwell's laws of electromagnetism, which we will study in the next section. The usual method of solving these equations is to introduce a function ϕ, called the *gravitational potential*, satisfying $\mathbf{G} = -\nabla\phi$. (This is possible since $\mathbf{G}$ is a conservative field by (12).) Note that ϕ is only defined up to a constant. (The difference between the values of ϕ at two points in space measures how much work must be done against gravity to move an object from one point to the other.) Substituting the expression for $\mathbf{G}$ into (15) we obtain *Poisson's equation*

$$\nabla^2\phi = 4\pi G\rho. \tag{16}$$

This is a *second-order* differential equation (it involves second derivatives), but it has the advantage of only involving a scalar function and not a vector field.

To summarize: to understand the gravitational field generated by a continuous distribution of matter, we only have to solve a second-order (scalar) partial differential equation (Poisson's equation).

Electricity and Magnetism

Electrostatics, the study of electrical forces between static or slowly moving charges, is formulated in mathematical terms in almost the same way as Newtonian gravitation. The key physical fact is *Coulomb's law*, which says the electrical force between two charged objects is an inverse square law. The only differences from the laws of gravity are that:

- the force between charges is proportional to the product of the charges, and does not involve the masses; and
- the force between charges can be either attractive or repulsive. Like charges repel; opposite charges attract.

With only changes in some constants, all the equations we derived for gravitation apply equally well to electrostatics. We spare the reader the chore of rederiving the results; we just state them. First, any distribution of charges in space defines an *electric field* $\mathbf{E}(\mathbf{r})$, a vector field with the property that the electrostatic force on a particle of arbitrary charge q with position vector $\mathbf{r}$ is given by $q\mathbf{E}(\mathbf{r})$. Second, neglecting magnetic effects, the electrical field $\mathbf{E}(\mathbf{r})$ obeys the analogues of (12) and (13):

$$\mathbf{curl}\,\mathbf{E} = 0 \tag{12'}$$

and *Gauss' law of electrostatics*:

$$\iint_S \mathbf{E} \cdot \mathbf{n}\, dS = 4\pi(\text{charge enclosed by } S). \tag{13'}$$

Since $\mathbf{E}$ is a conservative force field, we can again represent it in terms of a potential function ϕ, now called the *electrostatic potential*:

$$\mathbf{E} = -\nabla \phi.$$

For a point charge q located at the origin in $\mathbf{R}^3$, the electrostatic potential is simply

$$\phi(\mathbf{r}) = \frac{q}{\|\mathbf{r}\|}.$$

In the case of continuous charge distributions, ϕ obeys an analogue of Poisson's equation (16):

$$\operatorname{div} \mathbf{E} = -\nabla^2 \phi = 4\pi\rho, \tag{16'}$$

where ρ is the charge density. Equations (13') and (16') are the simplest forms of *Maxwell's equations*, which codify the principles of electromagnetism in terms of vector calculus.

One phenomenon that appears in electrostatics but not in gravitation is the concept of a (perfect) *conductor*. In a perfect conductor, charges instantaneously rearrange themselves so that the electrical field inside the conductor vanishes. Since $\mathbf{E} = -\nabla \phi$, the electrostatic potential ϕ is constant throughout the conductor. In practice, metal objects can usually be treated as conductors for purposes of electrostatics, as can moist earth. (Hence the term "to ground" a piece of equipment, i.e., to connect it to a perfect conductor so as to set $\phi = 0$ on it.) At the opposite extreme from a perfect conductor is a *perfect insulator* or *dielectric* (from the prefix *dia-*, meaning "apart"), inside of which current is unable to flow.

Thanks to the work of Maxwell, we know that it is impossible to give a satisfactory mathematical treatment of electrical forces without considering magnetism at the same time. Indeed, electricity and magnetism are two different aspects of the same fundamental force, usually called *electromagnetism*. Maxwell's equations of electromagnetism are coupled equations for two vector fields, the electric field $\mathbf{E}(\mathbf{r})$ and the magnetic field $\mathbf{B}(\mathbf{r})$. Only in static situations (where all charges are at rest) do these fields become decoupled from one another. The magnetic field, like the electric field, exerts a force on charged objects, but this force is proportional to the speed of the charged object and is perpendicular to its motion. The magnetic force on a charged object with charge q and position vector $\mathbf{r}$ is

$$\mathbf{F}_{\text{mag}} = \frac{-q}{c} \mathbf{B} \times \dot{\mathbf{r}},$$

where c is a universal constant, the speed of light in a vacuum. In a dynamic situation, Maxwell's equations are

$$\mathbf{curl\,E} = -\frac{1}{c}\dot{\mathbf{B}},$$

$$\mathrm{div\,E} = 4\pi\rho,$$

$$\mathbf{curl\,B} = \frac{4\pi}{c}\mathbf{J} + \frac{1}{c}\dot{\mathbf{E}}, \tag{17}$$

$$\mathrm{div\,B} = 0.$$

Here **J** is the *current density* (current per cross-sectional area). The field **J** has the same units as $\dot{\mathbf{E}}$, electric field per unit time. The equation for **curl E** is a form of *Faraday's law of induction*, which says that moving charges give rise to a magnetic field. The equation for **curl B** is a form of *Ampère's law*, which says that magnetic fields exert forces on moving charges. The equation for div E we recognize as a form of Gauss' law. The vanishing of div B means that there are no point sources of magnetic field, that is, no *magnetic monopoles*. (While physical theories involving magnetic monopoles, first proposed by Dirac in 1931, are increasingly in vogue, there is no firm experimental evidence that magnetic monopoles exist in nature.) Since div B $= 0$, it follows that we can find a vector field **A** (only unique up to addition of the gradient of a function) such that **B** = **curl A**. Such a field is called a *vector potential*.

Fluid Flow

In this section, we discuss the study of fluids using vector calculus. Fluid flow is a very complicated subject with applications ranging from weather prediction to the design of chemical factories. We can only give the most rudimentary introduction here. This should be enough to convince you of the importance of multivariable calculus for this subject. We consider a fluid (a liquid or a gas, basically "anything that flows") moving in a certain region $\mathcal{R}$ (possibly a container or the interior of a pipe) located in $\mathbf{R}^3$. At a point of $\mathcal{R}$ with position vector **r**, we can measure various properties of the fluid at that point, such as the temperature T, the pressure p, the density ρ, and the velocity vector **v**. The temperature, pressure, and density are related by the *equation of state* of the fluid. For example, for an "ideal gas," p is proportional to ρT. For liquids, on the other hand, ρ tends to vary very little with p and T. We will mostly be concerned with an idealized *incompressible fluid*, in which ρ is constant. We will also neglect thermodynamic effects and ignore the parameter T, though it is especially important in gas dynamics.

The first basic principle of fluid flow is *conservation of mass*, in other words, that fluid is neither created nor destroyed. That is not to say, however, that it can't flow from place to place. Consider a fixed closed surface S in $\mathcal{R}$. The flux integral $\iint_S \rho\mathbf{v} \cdot \mathbf{n}\,dS$ is the integral over S of the mass flowing across the surface per unit surface area per unit time. It therefore computes the mass of fluid flowing out of S per unit time. As the mass of fluid enclosed by S is

$$\iiint_{\text{interior of } S} \rho \, dV,$$

we obtain the equation

$$\iint_S \rho \mathbf{v} \cdot \mathbf{n} \, dS = \frac{\partial}{\partial t} \iiint_{\text{interior of } S} \rho \, dV.$$

Applying the divergence theorem to the left-hand side gives

$$\iiint_{\text{interior of } S} \operatorname{div}(\rho \mathbf{v}) \, dV = \iiint_{\text{interior of } S} \frac{\partial \rho}{\partial t} \, dV.$$

Since S was arbitrary, this gives us the equation of conservation of mass:

$$\operatorname{div}(\rho \mathbf{v}) = \frac{\partial \rho}{\partial t}. \tag{18}$$

In the case of an incompressible fluid, ρ is constant. So this simplifies to

$$\operatorname{div} \mathbf{v} = 0, \tag{18'}$$

which is analogous to equations (15) and (16′) in empty and uncharged space, respectively. By the way, this illustrates an important principle of mathematical physics: *very often, totally different physical phenomena are governed by the same mathematical equations.* That is one reason why vector calculus is so powerful.

The next equation of fluid flow corresponds to Newton's law of motion, *force = mass times acceleration*, applied to a small piece of the fluid. But we have to be careful. We've set up our coordinates to remain fixed in space, not to move with the fluid. For simplicity, suppose we're dealing with *steady flow*, so that $\mathbf{v}(\mathbf{r})$ does not vary with time. For a small piece of fluid located at the point with position vector $\mathbf{r}$ at time t, its velocity at time $t + h$ is $\mathbf{v}$ computed at the point to which the piece of fluid has moved by that time, which is (for small h) approximately $\mathbf{v}(\mathbf{r} + h\mathbf{v}(\mathbf{r}))$. By the chain rule, the acceleration of the piece of fluid is

$$\lim_{h \to 0+} \frac{\mathbf{v}(\mathbf{r} + h\mathbf{v}(\mathbf{r})) - \mathbf{v}(\mathbf{r})}{h} = \nabla \mathbf{v} \cdot \mathbf{v}.$$

(Since $\mathbf{v}$ is a vector, what we've denoted $\nabla \mathbf{v}$ is really the 3×3 matrix made up of the gradients of the three components of $\mathbf{v}$. Each of these gradients, when dotted with $\mathbf{v}$, gives one of the components of the acceleration.) The second equation of motion for steady flow is therefore of the form

$$\rho \nabla \mathbf{v} \cdot \mathbf{v} = \text{force on the fluid per unit volume.} \tag{19}$$

Equation (19) can become quite complicated when gravitational, electromagnetic, and thermodynamic effects have to be taken into account. In addition, we need to include *viscous forces*, forces due to internal friction in the liquid. The simplest case of (19) arises when the flow is steady, and we ignore all forces except those due to pressure in the fluid. (Pressure, after all, is force per unit area.) Then (19) simplifies to

$$\rho \nabla \mathbf{v} \cdot \mathbf{v} = -\nabla p. \tag{20}$$

(The "minus sign" reflects the fact that if the pressure increases in a certain direction, then that will tend to push fluid backward, from the region of high pressure to the region of low pressure.)

Since we usually don't know the variation of the pressure in the fluid, (20) might not appear to be of much use. But at least ∇p is a conservative vector field, so the curl of the left-hand side must vanish. Thus, in the very simplest case of fluid flow (steady incompressible flow with no viscous or external forces), we get a pair of coupled equations for the velocity field: (18′) and

$$\mathbf{curl}\,(\nabla \mathbf{v} \cdot \mathbf{v}) = 0. \tag{21}$$

Equation (21) is still a formidable equation to solve. However, you can check that it is satisfied when

$$\mathbf{curl}\,\mathbf{v} = 0. \tag{22}$$

This is the case of *irrotational flow*, and is a reasonable approximation to the truth when certain thick liquids are flowing rather slowly. Some cases of steady, incompressible, irrotational flow are studied in the following problem set. Note that the pair of equations (18′) and (22) coincides with some special cases of Maxwell's equations (17) of electromagnetism.

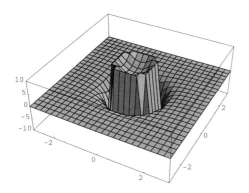

Problem Set I

PHYSICAL APPLICATIONS

1. Here's a famous problem in electrostatics. The objective is to compute the electrostatic potential and electrical field in the half-space $z \geq 0$ of $\mathbf{R}^3$ due to a point charge Q located at the point $(0, 0, 1)$ (i.e., one unit above the x-y plane), assuming the x-y plane is an infinite conducting plate. This gives a crude idealized model for the electrical field in a thunderstorm: the x-y plane represents the ground and the point $(0, 0, 1)$ represents the location of a charged thundercloud.

   First observe that the electrostatic potential and electrical field are the same as those due to *two* point charges, Q at $(0, 0, 1)$ and $-Q$ at $(0, 0, -1)$. Why is this? Add the potentials due to the two charges (separately), and differentiate the potential to compute the electric field. We will look at the cross sections of these fields in the x-z plane. Use the *Mathematica* commands `ContourPlot` and `PlotVectorField` to plot the electrostatic potential and electrical field, respectively, in the portion of the x-z plane with $-1 \leq x \leq 1$, $0 \leq z \leq 2$, when $Q = 1$. Combine the two plots on the same graph. (Hint: You may encounter difficulty with a nonnegative denominator function that vanishes at some points. Here is a trick to overcome that. If $f(x) = g(x)/h(x)$, $h(x) \geq 0$, set

   ```
   f1[x_] = If[h[x] > 0, f[x], 0]
   ```

   You will have to do a similar thing in a multivariable situation.)

2. Here is another problem in electrostatics. Compute and draw a contour plot of the electrostatic potential ϕ in $\mathbf{R}^3$ due to two parallel conducting circular loops which are oppositely charged (with charges of the same absolute value). For convenience we'll assume the charged loops are located at the circles $\{x^2 + y^2 = 1, z = 1\}$ and $\{x^2 + y^2 = 1, z = -1\}$. This is a crude model for a *capacitor*, a circuit element consisting of two oppositely charged conductors separated by a small nonconducting gap. Here is how to do the calculation. In the last problem, you computed the electrostatic potential at a point $\mathbf{r}$ in $\mathbf{R}^3$ due to a charge of $+1$ at $(0, 0, 1)$ and a charge of -1 at $(0, 0, -1)$. Simply by shifting, this gives the potential at $\mathbf{r}$ due to a charge of $+1$ at $(\cos \theta, \sin \theta, 1)$ and a charge of -1 at $(\cos \theta, \sin \theta, -1)$. You then want to integrate this potential over θ as θ runs from 0 to 2π. (By circular symmetry, the charge is uniform in θ.) In fact, *Mathematica* can do the integral to compute $\phi(x, 0, z)$. First compute the integrand as a function of x, z, and θ, by replacing (x, y, z) in the potential field of Problem 1 by $(x - \cos \theta, \sin \theta, z)$, and applying `Simplify`. (Once we compute ϕ at $(x, 0, z)$, symmetry around the z-axis implies that its value at (x, y, z) is $\phi(\sqrt{x^2 + y^2}, 0, z)$.) Now do the integral. The answer will be in terms of certain special functions. *Mathematica* can handle them. Draw contour plots of the solution for ϕ in the regions $0 \le x \le 4, 0 \le z \le 4$ and $-4 \le x \le 4, -4 \le z \le 4$. Then differentiate the potential function to compute the electric field. Do not be intimidated by the length and complexity of the answers. Plot the field in the first quadrant of the x-z plane using `PlotVectorField`. (Remember to load it first with the command `<<Graphics'PlotField'`.) You can save yourself some error messages if you use the region $0.01 \le x \le 4, 0.01 \le z \le 4$, and you can improve the picture by using the `ScaleFunction` option. Be patient with these plots. Incidentally, you can improve the plots further by increasing the number of `Contours` and `PlotPoints` in `ContourPlot`. Also, try adding color.

3. For applications in astrophysics, it is often important to understand the gravitational fields due to various mass distributions. As you have seen in Chapter 9, there are two main equations for studying such fields: Gauss' law and Poisson's equation. In this problem, we will study both of these equations in the case of mass distributions with cylindrical symmetry.

   (a) Suppose the mass density in space, ρ, is a function only of the radial coordinate r in cylindrical coordinates. Then the gravitational field $\mathbf{G}$ must also have cylindrical symmetry, and so (since the force is attractive) points straight inward toward the z-axis. Thus $\mathbf{G}$ is of the form $-g(r)(x\mathbf{i} + y\mathbf{j})/r$, where $g(r)$ is the field strength. (Here $-(x\mathbf{i} + y\mathbf{j})/r$ is a unit vector pointing toward the z-axis.) Let us apply Gauss' law to a right circular cylinder Σ centered along the z-axis, with height h and radius a. Since **G** points toward the z-axis, there is no flux through the top and bottom of the cylinder. Thus Gauss' law becomes

flux through $\Sigma = $ (area of sides of Σ) $\cdot (-g(a)) = -4\pi G$(mass enclosed by Σ),

or

$$(2\pi ah)g(a) = 4\pi G \int_0^{2\pi} \int_0^h \int_0^a \rho(r) r \, dr \, dz \, d\theta.$$

(Note the integration factor in cylindrical coordinates.) Compute the θ and z integrals by sight; then solve for $g(a)$ in terms of $\int_0^a \rho(r) r \, dr$. Then specialize in the case of the following density functions:

(i) $\rho(r) = \begin{cases} \rho \text{ (a constant)}, & r \leq 1, \\ 0, & r > 1. \end{cases}$

(ii) $\rho(r) = \dfrac{1}{(1 + r^2)^2}.$

(iii) $\rho(r) = e^{-r^2}.$

(iv) $\rho(r) = r^2 e^{-r^2}.$

Evaluate the integrals to compute $g(r)$ as a function of r, using $G = 1$. Plot the results. What happens to the field strength $g(r)$ as $r \to \infty$?

(b) Recall from Chapter 9 that the gravitational field **G** is given by $\mathbf{G} = -\nabla\phi$, where ϕ (the gravitational potential) satisfies Poisson's equation

$$\nabla^2\phi = 4\pi G\rho.$$

If ρ is a function of r only, then the same will be true for ϕ, again by symmetry considerations. Load the `Calculus 'VectorAnalysis'` package and use the command `SetCoordinates[Cylindrical[r, theta, z]]` to calculate in cylindrical coordinates. You can represent the left-hand side of Poisson's equation by the *Mathematica* expression `Laplacian[phi[r]]`. For each of the mass distributions (i)–(iv) of (a), solve Poisson's equation using `DSolve` and verify that you get the same formulas for the field strength function as in (a). You may take $G = 1$. (Notes: The potential functions may involve some functions you haven't seen before. If so, you can learn about them using the *Mathematica* Help Browser. Also, the solution to a second-order differential equation involves two constants of integration. One of these will be constrained by the requirement that ϕ not blow up as $r \to 0_+$. You can use `Series[phi[r], {r, 0, 2}]` to compute the first few terms of the power series of the solution function `phi[r]`, and then set the coefficients of $\log r$ terms or of negative powers of r to zero. This will still leave one constant of integration undetermined, but it will go away when you take the gradient to compute the field strength.)

4. This problem is almost identical to the last one, except that we will investigate mass distributions with spherical symmetry instead of those with radial symmetry. Take r to be the radial coordinate in *spherical coordinates*, and let Σ be a sphere centered at the origin with radius a. We again assume the mass density ρ is a function of r alone. This time the gravitational field takes the form $\mathbf{G} = -g(r)(x\mathbf{i} + y\mathbf{j} + z\mathbf{k})/r$, and Gauss' law becomes

$$\text{flux through } \Sigma = (\text{area of } \Sigma) \cdot (-g(a)) = -4\pi G(\text{mass enclosed by } \Sigma),$$

or

$$(4\pi a^2)g(a) = 4\pi G \int_0^{2\pi} \int_0^{\pi} \int_0^a \rho(r) r^2 \, dr \, \sin\phi \, d\phi \, d\theta.$$

Redo all the calculations of parts (a) and (b) of the last problem for cases (i)–(iv). This time you will need to use the command

```
SetCoordinates[Spherical[r, phi, theta]].
```

(In part (b), name the potential function something other than `phi` to avoid conflict of notation.) Are there any differences between your results this time and the cylindrical case from Problem 3? Pay attention to the rates of decay at infinity.

5. The most fundamental mathematical problem in both gravitation and electrostatics is the solution of *Laplace's equation* $\nabla^2\phi = 0$ for the potential function. (This equation also crops up in problems about fluid flow and heat flow.) Hence, it is of particular importance to have good methods for solving this equation with various boundary conditions. An important special case is the *Dirichlet problem*, where $\nabla^2\phi = 0$ is to be satisfied by a function ϕ defined in a region $\mathcal{R}$, and ϕ is specified *a priori* on the boundary of the region. In this case, it is known under reasonable conditions (see the Appendix below) that the solution is the function ϕ with the specified boundary values for which the energy $\iiint \|\nabla\phi\|^2 \, dV$ is minimized. This suggests the following numerical algorithm for solving the Dirichlet problem: (1) Start with some reasonable function ϕ_0 defined in $\mathcal{R}$ with the given values on the boundary. (This function will not usually satisfy Laplace's equation.) (2) Find a general class of functions ϕ_1 defined in $\mathcal{R}$ which are relatively "tractable" and which *vanish* on the boundary of $\mathcal{R}$. (3) Find the function $\phi = \phi_0 + \phi_1$ for which the energy integral is minimized (among all possibilities for ϕ_1). This should be a good approximation to a solution.

Now let's carry out this algorithm in a simple case. For convenience, we consider a two-dimensional instead of a three-dimensional situation; the mathematics is similar but the calculations are less complicated. Say we want to solve $\nabla^2\phi = 0$ in the interior of the square $\{0 \le x \le 1, 0 \le y \le 1\}$, and we're given that ϕ vanishes on the lower three sides of the square and is given on the top side $\{0 \leq x \leq 1, y = 1\}$ by $\phi(x, 1) = x(1 - x)$. A function ϕ_0 satisfying these boundary conditions is $\phi_0(x, y) = yx(1 - x)$. (Check this.) For a function ϕ_1 vanishing on the sides of the square, we can use

$$\phi_1(x, y) = x(1 - x)y(1 - y)(\text{any polynomial in } x, y).$$

Let the factor in parentheses be the most general polynomial in two variables of total degree 3. Then compute the energy integral (as a function of the coordinates) and search for a minimum using `FindMinimum`. You should not expect a simple form for the minimizing function. Draw a contour plot of your approximate solution for ϕ, and compare it with a contour plot of ϕ_0 (which does not satisfy Laplace's equation). Interpret the difference between the two plots.

6. In this problem we use the discussion of Newtonian gravitation in Chapter 9 to study the gravitational field of the Earth (or other planets).

(a) According to equation (14) in Chapter 9, the gravitational field at the surface of a spherically symmetric spherical body points in toward the center of the object, with field strength $g = GM/R^2$, where G is the universal gravitational constant, M is the mass of the object, and R is the radius. Compute the mass and density of the Earth (to three significant digits), given that $G = 6.674 \times 10^{-8} \, \text{cm}^3/(\text{g} \cdot \text{sec}^2)$, the gravitational field g at the surface of the Earth has field strength $978.0 \, \text{cm/sec}^2$, and the radius R of the Earth is about $6.37 \times 10^8 \, \text{cm}$. Do the same calculation for the Moon, which has radius about $1.74 \times 10^8 \, \text{cm}$ and gravitational acceleration about $162.7 \, \text{cm/sec}^2$. What might account for the difference in densities? (If you don't know, you can find the answer, as well as some additional information relevant to the next part of the problem, in a book on planetary science such as *Worlds Apart: A Textbook in Planetary Sciences*, by G. Consolmagno, S. J., and M. W. Schaefer (Prentice Hall, Englewood Cliffs, New Jersey, 1994).)

(b) The gravitational field of the Earth is of considerable importance for such purposes as rocket and satellite navigation, and it's known to an extraordinary degree of precision. There are variations from the model of a perfectly symmetrical planet due to a variety of factors: inhomogeneity of the Earth's interior, variations in surface topography, and above all, flattening of the Earth at the poles due to the centrifugal force coming from the Earth's rotation. In addition, centrifugal force itself exerts an effect on objects on the Earth's surface which is indistinguishable from weak *negative gravity* (pointing in the opposite direction from true gravity and varying quite considerably with latitude). In the rest of this problem we will explore some of these factors.

To a reasonable degree of accuracy, the shape of the Earth is a spheroid (i.e., an ellipsoid with radial symmetry about its axis of rotation), with equatorial radius $a = 6.378139 \times 10^8$cm and polar radius c, where the flattening constant is

$$\frac{a - c}{a} = 3.35282 \times 10^{-3}.$$

Our aim in the rest of the problem is to compute the gravitational field strength at the surface of such a spheroid in terms of GM, a, c, and the latitude φ. (Be sure to distinguish between the potential ϕ and the latitude φ. Recall that $\varphi = \pi/2$ corresponds to the North Pole, $\varphi = 0$ corresponds to the equator, and $\varphi = -\pi/2$ corresponds to the South Pole.) You may assume for purposes of this calculation that the density is constant. Here's a hint as to how to proceed. If $a = c$, the spheroid is a sphere and everything is spherically symmetric. In this case, we know from formula (14) in Chapter 9 that the gravitational field strength at a distance $r \leq a$ from the center of the object is

$$\frac{G}{r^2}(\text{mass of a sphere of radius } r) = \frac{G}{r^2}\frac{r^3}{a^3} \cdot M = \frac{GMr}{a^3}.$$

The gravitational potential $\phi_{\text{grav}}(r)$ may therefore be taken to be

$$\int_0^r \frac{GMx}{a^3}\,dx = \frac{GMr^2}{2a^3} = \frac{GM}{2a^3}\left(x^2 + y^2 + z^2\right), \qquad r \leq a,$$

and this satisfies Poisson's equation (equation (16) in Chapter 9) with $\rho = 3M/(4\pi a^3)$, i.e.,

$$\nabla^2\phi_{\text{grav}} = 4\pi G\rho = \frac{3GM}{a^3}, \qquad r \leq a.$$

Now the spheroid

$$\frac{\left(x^2 + y^2\right)}{a^2} + \frac{z^2}{c^2} = 1 \qquad (*)$$

is the image of the sphere $x^2 + y^2 + z^2 = a^2$ under the linear change of coordinates

$$x \mapsto x, \qquad y \mapsto y, \qquad z \mapsto \frac{cz}{a},$$

which has Jacobian

$$\det\begin{pmatrix} 1 & 0 & 0 \\ 0 & 1 & 0 \\ 0 & 0 & c/a \end{pmatrix} = \frac{c}{a};$$

so the volume of the spheroid is

$$\frac{c}{a}\frac{4}{3}\pi a^3 = \frac{4}{3}\pi a^2 c.$$

The *gravitational potential* ϕ_{grav} for a spheroidal object of constant density and mass M should therefore satisfy

$$\nabla^2\phi_{\text{grav}} = 4\pi G\rho = \frac{3GM}{a^2 c} \qquad (\dagger)$$

in the interior of the object.

However, on a rotating planet, the *effective potential* is

$$\phi_{\text{eff}} = \phi_{\text{grav}} + \phi_{\text{cent}},$$

where ϕ_{cent} is the *centrifugal potential*, i.e., $-\nabla\phi_{\text{cent}}$ is the centrifugal accelera-tion. Compute ϕ_{cent}, assuming that the spheroid $(*)$ is rotating about the z-axis with angular velocity ω. The potential ϕ_{cent} should be a constant multiple of $x^2 + y^2$.

(c) Finally, to compute ϕ_{grav}, we use the principle that the effective potential

$$\phi_{\text{eff}} \text{ should be constant on the surface of the object.} \tag{$\ddagger$}$$

This condition corresponds to the physical principle that gravity and centrifugal forces should not cause the oceans to flow from one part of the Earth to another, but that instead, *sea level* should be a surface of constant (effective) potential. Thus the solution for ϕ_{eff} should be of the form

$$\phi_{\text{eff}}(x,\, y,\, z) = C\left(x^2 + y^2 + \left(\frac{az}{c}\right)^2\right),$$

which ensures that ($\ddagger$) is automatically satisfied, and that the effective gravita-tional force $\mathbf{G}_{\text{eff}} = -\nabla\phi_{\text{eff}}$ is normal to the surface of the planet. Compute—in terms of c, G, M, and a—the correct value of C. To do this, subtract off ϕ_{cent}, which you already computed, and adjust C to satisfy (†). Then take the gradient to compute the effective gravitational field $\mathbf{G}_{\text{eff}}$. Compute the field strength as a function of φ by putting $\mathbf{r} = (a\cos\varphi,\, 0,\, c\sin\varphi)$, and then evaluat-ing $g = \|\mathbf{G}(\mathbf{r})\|$, φ the latitude along the spheroid $(*)$. Note that the effective gravitational field *does not always point toward the center of the object* in the spheroidal case. Compute what you get numerically for $g(\varphi)$ using the parame-ters of the Earth given above. Substitute the values of a and c given earlier, take $GM = 3.98603 \times 10^{20}\,\text{cm}^3/\text{sec}^2$, and take

$$\omega = \frac{2\pi}{24\,\text{hr}} = \frac{\pi}{12 \cdot 60^2\,\text{sec}}.$$

Compare your result with the internationally accepted *reference gravity field* of the Earth, which in units of cm/sec^2 is

$$g(\varphi) = 978.031846\left(1 + 0.005278895\sin^2\varphi + 0.000023462\sin^4\varphi\right).$$

7. Maxwell's equations imply that an electric current induces a magnetic field. Suppose for simplicity that a steady-state current J is flowing in a loop around the unit circle $x^2 + y^2 = 1$ in the x-y plane. The *Biot-Savart law* computes the

induced magnetic field at a point in $\mathbf{R}^3$ with position vector $\mathbf{r}$ to be

$$\mathbf{B}(\mathbf{r}) = \frac{J}{c} \int_0^{2\pi} \frac{d}{d\theta}(\cos\theta, \sin\theta, 0) \times \frac{\mathbf{r} - (\cos\theta, \sin\theta, 0)}{\|\mathbf{r} - (\cos\theta, \sin\theta, 0)\|^3} \, d\theta,$$

where c is the speed of light. Because of the rotational symmetry of the problem around the z-axis, it is only necessary to compute this for $\mathbf{r} = x\mathbf{i} + z\mathbf{k}$ in the x-z plane. The numerator in the integrand then becomes

$$(-\sin\theta\,\mathbf{i} + \cos\theta\,\mathbf{j}) \times (\mathbf{r} - \cos\theta\,\mathbf{i} - \sin\theta\,\mathbf{j}) \, d\theta$$
$$= \big(-\sin\theta\,\mathbf{i} \times (x\mathbf{i} + z\mathbf{k} - \cos\theta\,\mathbf{i} - \sin\theta\,\mathbf{j})$$
$$+ \cos\theta\,\mathbf{j} \times (x\mathbf{i} + z\mathbf{k} - \cos\theta\,\mathbf{i} - \sin\theta\,\mathbf{j}) \big) \, d\theta$$
$$= \big(z\sin\theta\,\mathbf{j} + \sin^2\theta\,\mathbf{k} - x\cos\theta\,\mathbf{k} + z\cos\theta\,\mathbf{i} + \cos^2\theta\,\mathbf{k} \big) \, d\theta$$
$$= \big(z\sin\theta\,\mathbf{j} + (1 - x\cos\theta)\mathbf{k} + z\cos\theta\,\mathbf{i} \big) \, d\theta.$$

The denominator is

$$\|\mathbf{r} - (\cos\theta, \sin\theta, 0)\|^3 = \big((x - \cos\theta)^2 + \sin^2\theta + z^2 \big)^{3/2}$$
$$= \big(x^2 - 2x\cos\theta + \cos^2\theta + \sin^2\theta + z^2 \big)^{3/2}$$
$$= \big(x^2 - 2x\cos\theta + 1 + z^2 \big)^{3/2},$$

which is a nonnegative even function of θ. The product of the denominator with $\sin\theta$ is an odd function of θ, and so the $\mathbf{j}$-term in the integral cancels out. Thus there are only two components of the integral to compute. Take the constant J/c to be 1.

(a) Use `Integrate` to find formulas for the $\mathbf{i}$ and $\mathbf{k}$ components `field1` and `field3` of the magnetic field as functions of x and z. Use the region $-3.01 \leq x \leq 3, -3.01 \leq z \leq 3$ to force the plotting routine to avoid the singularities of the field. Then use the command `PlotVectorField` to view the magnetic field in the x-z plane. Also, plot with `Plot3D` the $\mathbf{k}$ component of the field (the only component which is nonzero). By rotational symmetry, the function to be plotted is `field3[Sqrt[x^2 + y^2], 0]`. Plot on the region $-3.01 \leq x \leq 3, -3.01 \leq y \leq 3$ (again to avoid singularities) and set `PlotPoints` to `25`. Try adjusting `PlotRange`.

(b) It is often convenient to visualize the magnetic field in terms of "lines of force." These are the apparent patterns that you get by placing iron filings on a sheet of paper in the magnetic field, as you might remember from high-school physics. More rigorously, the lines of force are the curves traced out by a point with position vector $\mathbf{r}$ satisfying the differential equation of motion

$$\dot{\mathbf{r}} = \mathbf{B}(\mathbf{r}).$$

Use your answer to (a) and `NDSolve` to compute the lines of magnetic force around the current loop, and then plot these using `ParametricPlot`. Here are a few hints on how to set this up. You have to be careful since the equations are quite complicated and take a long time to solve, even on a fast computer. Please be patient! The commands

```
traj[t_, x0_] := {x[t], z[t]} /.  NDSolve[
{x'[t] == field1[x[t], z[t]], z'[t] == field3[x[t], z[t]],
x[0] == x0, z[0] == 0}, {x, z}, {t, 3},
MaxSteps -> 1250]
```

define a function `traj[t, x0]` that computes the line of force in the x-z plane through the point $(x_0, 0, 0)$ as a parametric function of t for $0 \leq t \leq 3$. The option `MaxSteps -> 1250` is designed to prevent the differential equation solver from terminating prematurely. (If the commands still take too long on your computer, you may need to settle for fewer trajectories.)

Make a `Table` of trajectories with $x_0 = (10j + 1)/60$, $0 \leq j \leq 5$. Use the symmetry of the problem to extend these solutions to more of the x-z plane, by reflecting the solutions across the x- and z-axes. Plot all the trajectories together on the same set of axes. (There should be 24 pairs of interpolating functions.) You should see a familiar pattern of nested loops of magnetic lines of force.

8. This problem deals with steady incompressible fluid flow, as discussed at the end of Chapter 9.

(a) Verify that equation (21) of Chapter 9, for steady incompressible fluid flow, is satisfied when **curl v** $= 0$ (the irrotational case). To do this, first check the identity

$$\nabla \mathbf{v} \cdot \mathbf{v} = (\mathbf{curl\ v}) \times \mathbf{v} + \tfrac{1}{2}\nabla(\mathbf{v} \cdot \mathbf{v}),$$

which holds for general vector fields **v**.

(b) Using the identity from (a), in the irrotational case where **curl v** $= 0$, deduce from equation (20) of Chapter 9 the principle known as *Bernoulli's law* for steady, irrotational, incompressible fluid flow; namely, $p + (\rho\|\mathbf{v}\|^2/2)$ remains constant.

(c) Suppose the equations div **v** $= 0$ and **curl v** $= 0$ for steady, irrotational, incompressible fluid flow are satisfied for a fluid contained in a region $\mathcal{R}$ of $\mathbf{R}^3$. Then, at least locally in $\mathcal{R}$, **v** is a conservative field and we can write $\mathbf{v} = \nabla\phi$ for some *potential function* ϕ. The condition div **v** $= 0$ then becomes *Laplace's equation* $\nabla^2\phi = 0$. Since fluid cannot flow through the sides of a closed container, we have the additional boundary condition $\partial v/\partial n = 0$ on the boundary of $\mathcal{R}$, or $\mathbf{n} \cdot \nabla\phi = 0$, where **n** is the outward unit normal. Show that if $\mathcal{R}$ is the half-space given by $x \geq 0$ in Cartesian coordinates (x, y, z), then $\phi(x, y, z) = y$ is a valid solution to the equations. Draw contours of the potential function and plot the velocity field for $0 \leq x \leq 3$ and $-3 \leq y \leq 3$. Superimpose the two plots to show the potential function and the velocity field simultaneously.

(d) Suppose $\mathcal{R}$ is the wedge given in cylindrical coordinates (r, θ, z) by $|\theta| \leq \pi/3$. Show that $\phi(r, \theta, z) = r^{3/2} \sin(3\theta/2)$ satisfies Laplace's equation and the boundary condition for steady, irrotational, incompressible fluid flow in $\mathcal{R}$. (You can use the command SetCoordinates to allow computation of the Laplacian in cylindrical coordinates.) Again, draw contours of the potential function and plot the velocity field, then superimpose the two. You can take $0 \leq x \leq 3$ and $-3 \leq y \leq 3$. (Hint: To block out the part of this rectangular region that is not physically relevant, use the command

```
black1 = Graphics[Polygon[{{0,0}, {Sqrt[3],3}, {0,3}}]]
```

to create a blackened triangle with vertices $(0, 0)$, $(\sqrt{3}, 3)$, and $(0, 3)$. Then superimpose it onto the plots with Show to block out the irrelevant part of the first quadrant. Similarly with the fourth quadrant.) Explain what is going on in the flow.

(e) In fluid flow problems, it is often useful to look at the *streamlines*, that is, the curves that would be traced by objects "drifting with the current." Since the streamlines must everywhere be tangent to the velocity field, they must be perpendicular to the level curves of the potential function ϕ. (Why? If necessary, refer back to Chapter 5 on *Directional Derivatives* and to Problem 2 in Problem Set E.) Show that the gradient of the function $h = r^{3/2} \cos(\frac{3}{2}\theta)$ is orthogonal to the gradient of ϕ, and thus that the streamlines are the level curves of h. Draw a ContourPlot of h (with ContourShading set to False) and superimpose it on your plots in (d) to visualize the streamlines for the flow.

9. Here is another problem on steady, incompressible, irrotational fluid flow. Consider such a flow in the half-space $x > 0$, but around an *obstacle* located at $\{0 \leq x \leq 1, y = 0\}$. (In this problem, we have assumed that the z-coordinate plays no role, so the problem is effectively two-dimensional.) Thus the region $\mathcal{R}$ where Laplace's equation is to be satisfied is

$$\mathcal{R} = \{(x, y) \in \mathbf{R}^2 : x > 0, \text{ and } x > 1 \text{ if } y = 0\}.$$

Show that $\phi(x, y) = \sqrt{r} \sin(\theta/2)$ satisfies Laplace's equation and the correct boundary condition for the potential function, provided that (r, θ) are polar coordinates for the points $(x^2 - y^2 - 1, 2xy)$ with $-\pi < \theta < \pi$. (First check that if $(x, y) \in \mathcal{R}$, then the point $(x^2 - y^2 - 1, 2xy)$ can't lie on the negative x-axis, so that we can indeed choose θ to lie in the range $-\pi < \theta < \pi$. Then check the equations. As in the last problem, draw contours of the potential function and plot the velocity field for $0.01 \leq x \leq 3$ and $-3 \leq y \leq 3$; then superimpose the two. (You may want to use `ScaleFunction` to adjust the lengths of the vectors.) As before, you can blacken out the obstacle by creating a blackened line segment with:

```
black = Graphics[[Thickness[0.03], Line[[{0, 0}, {1, 0}]]]]
```

Explain what is going on in the flow. As in the last problem, show that the streamlines are the level curves of $g(x, y) = \sqrt{r}\cos(\theta/2)$, and plot them using a contour plot of the function g. Combine the streamlines with the vector field in a single picture.

Glossary of Some Useful *Mathematica* Objects

Commands

`ArcTan` With one argument, the inverse tangent function. With two arguments, computes the θ of polar coordinates.

`Curl` Computes the curl of a vector field.

`Div` Computes the divergence of a vector field.

`DSolve` Symbolic ordinary differential equation solver.

`FindMinimum` Finds a minimum value of a function of one or more variables.

`Grad` Computes the gradient of a function.

`Graphics` Creates a graphics object.

`ImplicitPlot` Plots curves that are given implicitly as $f(x, y) = 0$.

`Laplacian` Computes the Laplacian of a function.

`Line` Graphics primitive for creating a line segment.

`NDSolve` Numerical differential equation solver.

`PlotGradientField` Plots a two-dimensional gradient field.

`PlotVectorField` Plots a two-dimensional vector field.

`Polygon` Graphics primitive for generating a polygon.

`SetCoordinates` Establishes name and type of coordinates for the functions in the `Calculus 'VectorAnalysis'` package.

Options and Directives

`ColorFunction` Used for setting colors in graphs.

`Compiled` Indicates whether an expression should be automatically compiled.

`DisplayFunction` Controls whether a graphics command displays its output; useful in suppressing extra output when using the `Show` command.

`MaxIterations` Controls the number of iterations *Mathematica* will allow when running a numerical routine (such as `FindRoot`).

`MaxSteps` Specifies the maximum number of steps in `NDSolve`.

`PlotLabel` Labels a plot.

`PlotPoints` Specifies the number of points *Mathematica* will sample when generating a plot.

`ScaleFunction` Adjusts lengths of arrows in vector field plots.

`Thickness` Graphics primitive for setting thickness of curves and lines.

`WorkingPrecision` Specifies the precision in a numerical algorithm (such as employed in `FindRoot`).

Packages

`Calculus`VectorAnalysis`` Needed for `Grad, Div, Curl, Laplacian`.

`Graphics`PlotField`` Needed for `PlotVectorField` or for `PlotGradientField`.

Appendix: Energy Minimization and Laplace's Equation

In this Appendix we briefly explain why the energy minimization approach to Laplace's equation, outlined and used in Problem 5, gives the correct answer. Consider the *Dirichlet problem* in a bounded region $\mathcal{R}$ in $\mathbf{R}^3$, which is to find a function ϕ defined in a region $\mathcal{R}$, where ϕ is given on the boundary of the region and ϕ solves the equation $\nabla^2\phi = 0$ inside $\mathcal{R}$. We take it for granted that there is indeed a unique solution. (This requires proof, which for "reasonable" regions $\mathcal{R}$ is supplied in courses on partial differential equations.) Call the solution ϕ_0. (This is not necessarily the same as the starting function ϕ_0 for the algorithm described in Problem 5. But if it were, the algorithm would immediately terminate and yield the correct answer.) Then a general function in $\mathcal{R}$ with the given boundary values is of the form $\phi = \phi_0 + \phi_1$, where ϕ_1 *vanishes* on the boundary. Let

$$E(\phi) = \iiint_{\mathcal{R}} \|\nabla \phi\|^2 \, dV.$$

We want to show that $E(\phi) \geq E(\phi_0)$, with equality only if $\phi = \phi_0$, i.e., only if $\phi_1 = 0$ everywhere. But

$$
\begin{aligned}
E(\phi) &= \iiint_{\mathcal{R}} \nabla \phi \cdot \nabla \phi \, dV \\
&= \iiint_{\mathcal{R}} (\nabla \phi_0 + \nabla \phi_1) \cdot (\nabla \phi_0 + \nabla \phi_1) \, dV \\
&= \iiint_{\mathcal{R}} \nabla \phi_0 \cdot \nabla \phi_0 \, dV + 2 \iiint_{\mathcal{R}} \nabla \phi_1 \cdot \nabla \phi_0 \, dV + \iiint_{\mathcal{R}} \nabla \phi_1 \cdot \nabla \phi_1 \, dV \\
&= E(\phi_0) + 2 \iiint_{\mathcal{R}} \nabla \phi_1 \cdot \nabla \phi_0 \, dV + \iiint_{\mathcal{R}} \|\nabla \phi_1\|^2 \, dV.
\end{aligned}
$$

The last term on the right is ≥ 0, and we can compute the second term using the Divergence Theorem, since by the product rule,

$$\mathrm{div}(\phi_1 \nabla \phi_0) = \nabla \phi_1 \cdot \nabla \phi_0 + \phi_1 \nabla^2 \phi_0,$$

and $\nabla^2 \phi_0 = 0$ by assumption. So

$$
\begin{aligned}
2 \iiint_{\mathcal{R}} \nabla \phi_1 \cdot \nabla \phi_0 \, dV &= 2 \iiint_{\mathcal{R}} \mathrm{div}(\phi_1 \nabla \phi_0) \, dV \\
&= 2 \iint_{\mathrm{boundary}\, \mathcal{R}} (\phi_1 \nabla \phi_0) \cdot \mathbf{n} \, dS \\
&= 0,
\end{aligned}
$$

since $\phi_1 = 0$ on the boundary of $\mathcal{R}$. Thus we have seen that

$$E(\phi) = E(\phi_0) + \iiint_{\mathcal{R}} \|\nabla \phi_1\|^2 \, dV,$$

which shows that

$$E(\phi) \geq E(\phi_0),$$

with equality only if $\nabla \phi_1 = 0$ everywhere. But this means ϕ_1 is a constant function, and since $\phi_1 = 0$ on the boundary of $\mathcal{R}$, $\phi_1 = 0$ everywhere.

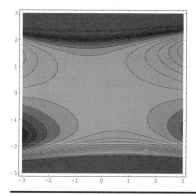

Chapter 10

MATHEMATICA TIPS

In this chapter, we will use a question-and-answer format to discuss common difficulties encountered by our students. Some of the difficulties are intrinsically mathematics issues; but most are deeply intertwined with students' attempts to apply *Mathematica* to solve mathematics problems.

Question 1. Why is *Mathematica* not doing anything? I entered a command, and nothing is happening!

Answer. First of all, make sure you typed SHIFT+ENTER at the end of the command, and not ENTER. Otherwise, *Mathematica* may simply be waiting for you to finish typing! Assuming you typed SHIFT+ENTER, *Mathematica* is not idle—it is probably churning away on a computation that is either very slow or that it cannot complete. While it is working, *Mathematica* will highlight the cell bracket of the input cell it is processing, and will type "Running ..." at the top of your Notebook. For instance, when you enter the first input line of a session, *Mathematica* takes a long time to respond because it is busy loading the main part of the program.

There is always a chance that some typing or syntax error has gotten *Mathematica* stuck in a process that will not terminate. If you feel that a response is overdue, click on the cell bracket that is highlighted, open the `Kernel` menu, and choose either `Interrupt Evaluation...` or `Abort Evaluation`. The former will give you a chance to suspend the calculation; the latter will terminate it.

Question 2. Why does `Solve` write its output in such a funny way, with arrows and two levels of braces? Wouldn't it be easier if it just printed out the answer?

Answer. On a first encounter, the output of `Solve` seems peculiar, especially in cases such as:

> In[1]:= `Solve[x + 1 == 0, x]`

> Out[1]= $\{\{x \rightarrow -1\}\}$

Here the outer level of braces seems superfluous, and we could get rid of it with the `Flatten` command. For equations with multiple solutions, *Mathematica* needs the outer level of braces in order to make the solutions into a *list*. The inner braces are used in case you solve for several variables, so they can again be grouped together in a list. Finally, the arrow is used because *Mathematica* reports each solution as a *transformation rule*. This conveniently allows you to substitute the solution into other expressions, using the *replacement operator*. Here's an example.

> In[2]:= `solns = N[Solve[x^3 + x + 1 == 0, x]]`

> Out[2]= $\{\{x \rightarrow -0.682328\}, \{x \rightarrow 0.341164 + 1.1615I\},$
> $\{x \rightarrow 0.341164 - 1.16154I\}\}$

Here the three (numerical) solutions of the cubic equation $x^3 + x + 1 = 0$ are presented as transformation rules, and the three rules make up a list. We can extract the real-valued solution by referring to it as the first element of the list. If we want the value of the derivative $3x^2 + 1$ at this solution, we can get it by typing

> In[3]:= `3x^2 + 1 /. solns[[1]]`

> Out[3]= 2.39671

The symbol `/.` denotes the *replacement operator*. It tells *Mathematica* to use the transformation rules (on the right) to replace variables by values in an expression (on the left). In this example, all occurrences of the variable **x** in the expression `3x^2 + 1` are replaced by the value -0.682328.

Question 3. I typed `f[x_] = x^2 + x + 1` and then `f[y]`. *Mathematica* responded with "3" instead of with $y^2 + y + 1$. How can that be?

Answer. Totally unexpected output is a common problem. In this case, the reason is almost certainly that, earlier in the same session, the variable **y** was set equal to 1. (The assignment could even have taken place in a different Notebook in the same session!) *Mathematica* never forgets. When something like this happens, clear the variables, for example, by typing `Clear[y]`. A more drastic solution is to quit and restart the kernel, or even to save your Notebook and exit and reenter *Mathematica*. (These methods clear all variables and allow you to start again.)

Question 4. I'm having trouble keeping straight the different kinds of brackets and braces that *Mathematica* uses. What do I use where?

Answer. You must use the three matched delimiters—*parentheses* ( ), *brackets* [ ], and *braces* { }—correctly. Braces are used only to denote lists. Brackets are used to encase arguments of functions and commands. Parentheses are used, as in mathematics, to separate expressions, variables, and constants whenever ambiguity is possible. Using the wrong delimiter can cause troublesome problems. The most common error is the misuse of parentheses where brackets are indicated. This error occurs because common mathematical functional notation calls for parentheses where *Mathematica* uses brackets. Here is a typically frustrating mistake:

```
In[4]:=  f[x_] = Cos(x); f[Pi] // N
```

```
Out[4]=  3.14159 Cos
```

Mathematica interpreted the function f[x] as the product of the variable x times the constant Cos. That explains the output. Similarly, *Mathematica* will complain if you try to enter lists with parentheses instead of braces:

```
In[5]:=  lis = (1, 2, 3)
      Syntax::sntxf:
        "(" cannot be followed by "1, 2, 3)".
      lis = (1, 2, 3)
```

Question 5. When I get a complicated algebraic answer, I realize I should have used `Simplify` in my command line. But then when I go back and try to put it in, I invariably mess up. Is there some way to compensate?

Answer. Yes. Use the double slash suffix operator. Any *Mathematica* command that takes a single argument can be entered this way. In place of

```
Simplify[(x^2 - y^2)/(x - y)]
```

you can type

```
(x^2 - y^2)/(x - y) // Simplify
```

The latter is more convenient when editing an input line.

Question 6. I edited an input line to add an option to a command. Now everything is broken. What went wrong?

Answer. Options must be placed after the required arguments to a command, but before the closing bracket. Here is an example:

```
In[6]:=  FindRoot[Exp[x] == x^10, {x, 10}]
```

> FindRoot::cvnwt: Newton's method
>   failed to converge to the prescribed
>   accuracy after 15 iterations.

```
Out[6]=  {x → 2.05971}
```

Mathematica has warned us that the answer may be inaccurate, so we want to repeat the calculation with more iterations. The following three attempts would all fail, because the option has been placed incorrectly.

```
FindRoot[Exp[x] == x^10, MaxIterations -> 30, {x, 10}]
FindRoot[Exp[x] == x^10, {x, 10}] MaxIterations -> 30
FindRoot[Exp[x] == x^10, {x, 10}] // MaxIterations -> 30
```

Here is the correct way to insert the option into the command.

```
In[7]:=  FindRoot[Exp[x] == x^10, {x, 10},
           MaxIterations -> 30]
```

```
Out[7]=  {x → 1.11833}
```

Question 7. I keep getting confused about the differences between =, :=, and ==. How can I tell which one to use?

Answer.   This is one of the trickiest aspects of *Mathematica* usage, but there are a few simple principles to keep in mind. The = operator makes an *immediate* and *permanent* assignment. This is useful for assigning shorthand names to expressions or constants you intend to use repeatedly. (However, a common error is to forget that a permanent assignment has been made. See Question 3.) The := operator also makes an assignment, but delays implementation until you actually invoke the name of the assigned quantity. Delayed evaluation can be useful in several situations. Typically, you might use a command like NDSolve with a variable in a place where the command requires a number. By delaying the evaluation, you can supply a number (perhaps still to be calculated) at a later time. Here is an example:

```
In[8]:=  f[c_] := NDSolve[{y'[x] == y[x]^2 + x, y[0] == c},
           y[x], {x, 0, 1}]
```

(Try omitting the colon in the definition of f to see what happens.) Each time you use f[c], you get the solution to the differential equation satisfying the initial condition $y(0) = c$. For example, to get the solution satisfying $y(0) = 1$, you would type

In[9]:= **f[1]**

Out[9]= $\{\{y[x] \rightarrow \text{InterpolatingFunction}[\{\{0., 1.\}\}, <>][x]\}\}$

Here is a further illustration of the difference between = and : =.

In[10]:= **Clear[f, g];**

 f[x_] = x*Random[ ];

 g[x_] := x*Random[ ]

In[11]:= **{f[1], f[1], f[2]}**

Out[11]= $\{0.330602, 0.330602, 0.661205\}$

In[12]:= **{g[1], g[1], g[2]}**

Out[12]= $\{0.549591, 0.420762, 0.207796\}$

The definitions of **f[x]** and of **g[x]** are almost identical; each is supposed to be **x** multiplied by a random number between 0 and 1. In the case of **f[x]**, immediate assignment is used, so the random number generator is invoked once and **f[x]** is permanently defined to be **x** times this random number (which turned out to be 0.330602). In the case of **g[x]**, however, the random number generator is invoked each time the function is called, and gives a different result each time.

Finally, == is used for setting up equations that are to be manipulated (as *equations* not as *assignments*). For instance, such equations can be solved by **Solve** or **FindRoot**. Applying **Solve** or **FindRoot** to an "equation" containing = in place of == will result in an error message, and may have the side effect of an unintended assignment. For example:

In[13]:= **Solve[x = 2, x]**

 General::ivar: 2 is not a valid variable.

Out[13]= $\text{Solve}[2, 2]$

In[14]:= **Solve[x == 2, x]**

 General::ivar: 2 is not a valid variable.

Out[14]= $\text{Solve}[\text{True}, 2]$

In In[13], *Mathematica* recognized that **x** = 2 is not a legitimate equation, but it still permanently replaced **x** by 2. Then, in In[14], it failed because the putative equation **x** == 2 is nothing more than the tautology 2 == 2. In order to proceed, you need to **Clear[x]**.

Question 8. Why doesn't *Mathematica* simplify expressions like $\sqrt{x^2}$? It should report it as x.

Answer. *Mathematica* errs on the side of caution. After all, it does not know whether x is positive or negative. The value of $\sqrt{x^2}$ is $\pm x$ accordingly. Of course, x could also be a complex number. In fact, *Mathematica*'s caution sometimes prevents it from simplifying expressions that it should be able to simplify. A typical example is

```
In[15]:= expr = x/Sqrt[x^4];
         Simplify[expr]
```

$$Out[15]= \frac{x}{\sqrt{x^4}}$$

The answer is $1/x$, regardless of the sign of the variable. Here is a trick that sometimes works; it assumes you are willing to live with positive answers.

```
In[16]:= Sqrt[Simplify[expr^2]]
```

$$Out[16]= \sqrt{\frac{1}{x^2}}$$

Question 9. When I try to plot, I sometimes get these awful error messages about "machine-size real numbers" and an empty plot. Why is this happening?

Answer. There are many possible reasons. We mention several below. Often, the reason is the omission of the **Evaluate** command. The difficulty arises from the precise sequence of events that transpires when *Mathematica* executes a plot command. For a detailed explanation, you can consult *The Mathematica Book*, 3rd ed., by S. Wolfram. Suffice it to say that, if you have any doubt, it never hurts to surround the expression you are intending to plot with **Evaluate**. This is particularly true if you use a command inside your plot call to generate the expression to be plotted. A typical example, illustrated below, occurs with the use of **Table** inside **Plot**.

```
In[17]:= Plot[Table[x^j, {j, 1, 2}], {x, 0, 2}]

         Plot::plnr : Table[x^j, {j, 1, 2}] is not a
         machine-size real number at
         x = 8.33333333333333214`*^-8.

         Plot::plnr : Table[x^j, {j, 1, 2}] is not a
         machine-size real number at
         x = 0.0811339831458315785`.

         Plot::plnr : Table[x^j, {j, 1, 2}] is not a
         machine-size real number at
         x = 0.169617599718747342`.
         General::"stop": "Further output of Plot :: plnr
         will be suppressed during this calculation.
```

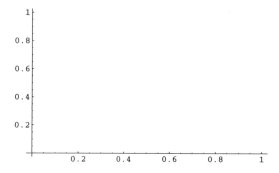

`Out[17]= —Graphics—`

Here is the correct syntax:

`In[18]:= ` **`Plot[Evaluate[Table[x^j, {j, 1, 2}]], {x, 0, 2}];`**

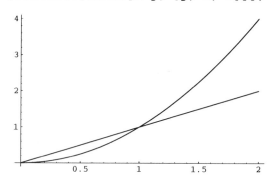

Here are a few other reasons for difficulties with plotting. You must enter a list of expressions of the size expected by the plotting command. For example,

`ParametricPlot[{Cos[x]}, {x, 0, Pi}]`

will not work because `ParametricPlot` expects a list of two expressions. Similarly, the command

`ListPlot[{{1, 2, 1}}]`

will fail because `ListPlot` plots points in the plane, not in space. And lastly, the command

`Plot[Exp[I*x], {x, 0, 1}]`

does not work because `Plot` requires real-valued expressions. In summary, be certain that you enter appropriate quantities into a plotting command.

Question 10. When I try to use `Show` to combine two or more plots, sometimes I see each of the plots before I see the combined plot. How can I prevent this?

Answer. It depends on whether your original plots were displayed or held in reserve. You can control whether a plot is displayed with the `DisplayFunction` option. For example, if you type

```
pic1 = Plot[x^2, {x, -1, 1}]
pic2 = Plot[3(1 - x^2), {x, -1, 1}]
Show[{pic1, pic2}]
```

as three successive input lines, you will see `pic1` after the first line, `pic2` after the second, and then they will be combined after the `Show` command. But if you type the three lines as

```
pic1 := Plot[x^2, {x, -1, 1}]
pic2 := Plot[3(1 - x^2), {x, -1, 1}]
Show[{pic1, pic2}]
```

then there will be no output from the first two lines, but after the third command, the two `pics` will display individually before the combined plot is displayed. You can prevent that by typing your lines as

```
pic1 := Plot[x^2, {x, -1, 1}, DisplayFunction -> Identity]
pic2 := Plot[3(1 - x^2), {x, -1, 1}, DisplayFunction ->
  Identity]
Show[{pic1, pic2}, DisplayFunction -> $DisplayFunction]
```

Then you will only see the single combined plot.

Question 11. The top of one of my plots seems to have been cut off. Another plot seems to include too much, so that the detail is washed out. What can I do about these plots?

Answer. *Mathematica*'s plotting routines are designed to look for the "interesting" part of a graph, but their judgment is not infallible. You can adjust the part of the plot that is shown with the `PlotRange` option. (You can also adjust the relative scales on the $x-$ and y-axes with the `AspectRatio` option. See the cycloid example in Chapter 3 and the torus and cylinder example in Chapter 8.) Here are some examples.

```
In[19]:= Plot[(x^2 - 1)^4 Sin[3x], {x, -3, 3}];
```

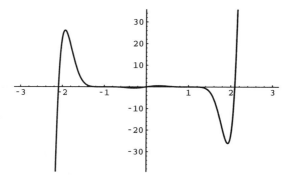

Here the "main features" of the graph are visible, but the detail for $-1 \le x \le 1$ is washed out, and it is not clear what happens for $|x| \ge 2.5$. To get more detail about the behavior of the function for small x, try restricting the **PlotRange**:

```
In[20]:= Show[%, PlotRange -> {-5, 5}];
```

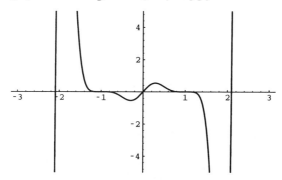

To get more information about the behavior of the function for large x, try showing the whole graph:

```
In[21]:= Show[%, PlotRange -> All];
```

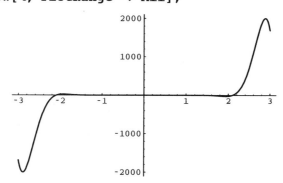

One extra caution regarding **Plot3D**: the default setting fills in parts of the surface where the function is outside the **PlotRange** and makes them look flat, so that the appearance of any flat portions of your plot might be quite misleading. You can turn off this feature with the option `ClipFill -> None`.

Question 12. I produced a contour plot of a function of two variables but it's quite jagged and doesn't seem very accurate. I thought *Mathematica*'s numerical routines were excellent. What can I do?

Answer. The problem probably lies not with the accuracy of *Mathematica*'s numerical routines, but rather with the way the plotting routines work. They evaluate the expression at a fixed number of points and then "connect the dots." You can improve the appearance of contour plots by increasing `PlotPoints`, which determines the number of points in each direction where *Mathematica* evaluates the expression. (The default for `ContourPlot` is 15.) Another possibility is to increase `Contours`, which controls how many (or which specific) contours to show. To illustrate, we provide an example where we first use the defaults, and another contour plot of the same function with different values of `PlotPoints` and `Contours`.

```
In[22]:= f[x_, y_] = x^3*y^2 - 47x^2 y - 33x*y^2 + 5y^7 - 3x;
```

```
In[23]:= ContourPlot[f[x, y], {x, -3, 3}, {y, -3, 3},
            ColorFunction -> Hue];
```

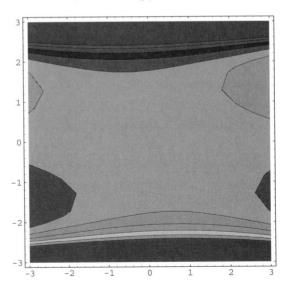

```
In[24]:= ContourPlot[f[x, y], {x, -3, 3}, {y, -3, 3},
            PlotPoints -> 40, ColorFunction -> Hue,
            Contours -> 30];
```

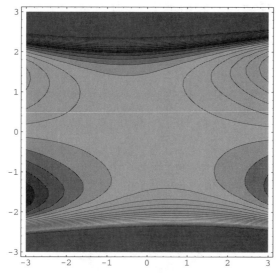

`PlotPoints` has the same meaning and default value in `Plot3D` that it has in `ContourPlot`, and you can again improve accuracy by increasing it.

Question 13. What's the difference between an `ImplicitPlot` of a level curve $f(x, y) = c$ and a `ContourPlot` of the expression $f(x, y)$ with the options `Contours -> {c}, ContourShading -> False`?

Answer. It depends on how you enter the `ImplicitPlot` command. If you give ranges for the values of both x and y, `ImplicitPlot` applied to `f[x, y] == c` is identical to `ContourPlot` applied to `f[x, y]` with the options `Contours -> {c}, ContourShading -> False, PlotPoints -> 25`. If you specify a range of values only for x, then `ImplicitPlot` works by choosing values of x and using `Solve` to find corresponding values of y. It then plots those points and connects them as in a `Plot` command. The second method is preferable when f is a polynomial function, since `Solve` works well on these. When f is a transcendental function, `Solve` can fail, and therefore the first method (using `ContourPlot`) is preferable.

Question 14. Why does `ImplicitPlot` sometimes give strange results?

Answer. This might be due to using the wrong method in `ImplicitPlot`. (See the previous question.) It can also be the result of a *mathematical* problem discussed in Chapter 5, *Directional Derivatives*. Namely, a level curve that passes through a critical point can be singular. This can lead to instabilities in the numerical algorithm. This is apparent in the following example:

```
In[25]:= <<Graphics`ImplicitPlot`
```

```
In[26]:= ImplicitPlot[Sin[x] - Cos[y] == 0, {x, -2Pi, 2Pi},
         {y, -2Pi, 2Pi}];
```

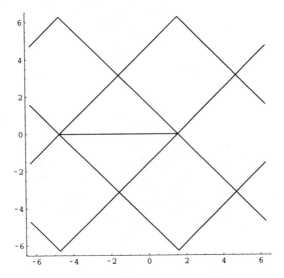

The correct solution to $\sin x - \cos y = 0$ comes from observing that $\sin x = \cos((\pi/2) - x)$. So, the equation becomes $\cos y = \cos((\pi/2) - x)$. Thus, $y = \pm((\pi/2) - x) + 2\pi n$ for some integer n. The correct plot should therefore be a collection of straight lines of slope ± 1. Note how the result changes when we replace 0 in the equation by the noncritical value 0.000001.

```
In[27]:= ImplicitPlot[Sin[x] - Cos[y] == 0.000001,
         {x, -2Pi, 2Pi}, {y, -2Pi, 2Pi}];
```

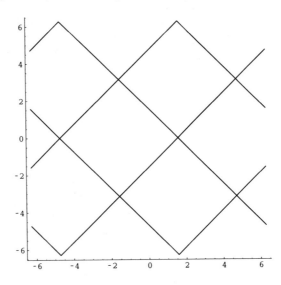

Question 15. I just opened a Notebook that was fine in an earlier session, but when I evaluated it, havoc ensued. What happened?

Answer. Here are the likely possibilities. First, you may have included input lines with references to other output lines via the percent sign, references like `%` or `%34`. It is possible that the line numbering has changed and the references are no longer accurate. Second, you may have used commands that live inside packages that are not loaded. That is a particularly nettlesome error as now you will have to `Remove` those commands before you load the package. Third, you need to be sure that your input lines are in the correct order. It is possible that in the prior session, you used some output in a new line that appears physically earlier in your Notebook. The order must be corrected. Finally, some of your definitions may have disappeared. This occurs when you open several Notebooks in a single session and you make a definition in one that applies to all. Unless you have the definition in the particular Notebook you have reopened, it will not be remembered from the last session.

Question 16. I'm having some trouble navigating the online help for *Mathematica*. What's the best strategy for using online help effectively?

Answer. The best way to start the Help Browser is to click on the `Help` button in the menu bar, and then click `Help....` There are six buttons at the top of the Help Browser. The button you select tells *Mathematica* where to search for help items. By default, the Help Browser appears with the "Built-in Functions" button selected. If you type the name of a built-in function, like `Sin`, next to the "Go To:" box and click the box, then information on it will appear in the main window below. *Caution*: It sometimes takes *Mathematica* a few seconds to locate the item in the database—even though the path immediately appears below the "Go To:" box. Resist the temptation to click on the items in the path; wait for information to appear in the main window. Often, there will be hyperlinks embedded in the information that will allow you to navigate the informational database.

If you want *Mathematica* to look specifically for information in "The Mathematica Book," select that button. We have found that selecting the "Master Index" button allows for the broadest searches. Another useful choice is the "Add-ons" button. You should select this button if you need information on commands or options that are found in packages. For example, to learn about `ImplicitPlot`, select the "Add-ons" button, then click "Standard Packages," then "Graphics," and finally "ImplicitPlot." If you did not know that `ImplicitPlot` was in the `Graphics` package, then you could search, under I, in the "Master Index."

Question 17. Where can I see more sample *Mathematica* notebooks and get additional information about *Mathematica*?

Answer. Check out the Wolfram web site at `http://www.wolfram.com`.

It contains lots of useful information, including a much longer lists of FAQs and other *Mathematica* tips.

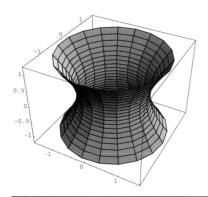

Glossary

This glossary is intended to supplement the chapter glossaries found throughout the book. In those, you found a roster of commands, options, and packages that were particularly useful in solving the problems from that chapter. In each case a brief description of the command or option was provided. In this more comprehensive glossary, we also include sample uses of the commands and options. We have included a few items that are not present in any of the chapter glossaries.

All of the examples are presented as input lines that use the specific command or option in question. The corresponding output lines are not reproduced. It will help your understanding if you enter the input and examine the resulting output. In the case of options, you should try the command both with and without the option. To make the chore easier, we have included all the input entries from this glossary on the diskette that accompanies this book.

Commands

`Arrow`   In the `Graphics'Arrow'` package.   Creates an arrow joining two indicated points in the plane.

```
Show[Graphics[Arrow[{0, 0}, {1, 0}]]];
```

`Chop` Replaces all real numbers in an expression, smaller than a prescribed value, by 0. Default is 10^{-10}. Useful in eliminating numerical "garbage" due to round-off errors.

```
expr = x + 0.0002x^2; Chop[expr, 0.01]
```

Circle Creates a circle with indicated center and radius.

```
Show[Graphics[Circle[{0, 0}, 1]],
    AspectRatio -> Automatic];
```

Clear Clears the definition of a variable.

```
x = 3; x
Clear[x]; x
```

ContourPlot Produces a two-dimensional plot of the level curves of a function $f(x, y)$.

```
ContourPlot[x^2 - y^2, {x, -1, 1}, {y, -1, 1}]
```

ContourPlot3D In the **Graphics`ContourPlot3D`** package. Produces a three-dimensional plot of the level surfaces of a function $f(x, y, z)$.

```
ContourPlot3D[x^2 + y^2 + z^2 - 1, {x, -1 , 1},
    {y, -1 , 1}, {z, -1 , 1}]
```

Cross Computes the cross product of two vectors in 3-space; if working with vectors in 2-space, add a zero third coordinate before invoking.

```
Cross[{2, 4, 6}, {1, 3, 5}]
c := {u, v}; d := {x, y};
Cross[Join[c, {0}], Join[d, {0}]]
```

CrossProduct In the **Calculus`VectorAnalysis`** package. Same effect as **Cross**, but also works in other coordinate systems.

```
c := {u, v}; d := {x, y};
CrossProduct[Join[c, {0}], Join[d, {0}]]
```

Curl In the **Calculus`VectorAnalysis`** package. Computes the curl of a vector field.

```
Curl[{x*y, x*z^2, y^2*z}]
```

CylindricalPlot3D In the **Graphics`ParametricPlot3D`** package. Plots surfaces given by parametric equations in cylindrical coordinates.

```
CylindricalPlot3D[r^2, {r, 0, 1}, {theta, 0, 2Pi}]
```

D Differentiates an expression. The symbol ∂ from the BasicInput palette has the same effect. One can also use the apostrophe **'**.

```
f[x_] := (x^4 + 2x^3 + 1)^3
D[f[x], x]
f'[x]
D[Sin[x^2 + y^3], y]
D[Tan[x^2], {x, 2}]
```

Det Computes the determinant of a square matrix.
```
Det[{{1, 2, 3}, {4, 5, 6}, {7, 8, 10}}]
```

Disk Creates a disk with indicated center and radius.
```
Show[Graphics[Disk[{0, 1}, 3]]]
```

Div Computes the divergence of a vector field.
```
Div[{x^2 + y^2, x, y}]
```

Do Repeats a command a specified number of times.
```
fact = 1; Do[fact = fact*i, {i, 10}]; fact
```

Dot Dot product. Also written with a period, as in A.B.
```
Dot[{-1, 2}, {3, 5}]
```

Drop Drops designated elements from a list. Useful in eliminating unwanted solutions found by Solve.
```
sollist = Solve[x^3 + x == 0, x]
Drop[sollist, {2, 3}]
```

DSolve Symbolic differential equation solver.
```
DSolve[y'[x] == y[x]^2 + 1, y[x], x]
DSolve[{y''[x] + (x - 1)*y'[x] + y[x] == 0,
    y[0] == 1, y'[0] == 0}, y[x], x]
DSolve[{x'[t] == y[t], y'[t] == -x[t] + y[t]},
    {x[t], y[t]}, t]
```

Eigenvalues Computes the eigenvalues of a square matrix.
```
Eigenvalues[{{1, 2, 3}, {4, 5, 6}, {7, 8, 10}}]
```

Eliminate Eliminates one or more variables from a set of equations.
```
Eliminate[{x^2 + y^2 == 3, x + y == 1}, x]
```

Evaluate Forces evaluation of an expression; useful when using commands inside a plotting routine.
```
Plot[Evaluate[Table[x^j, {j, 0, 3}]], {x, 0, 2}]
```

Expand Expands a complex algebraic expression.
```
Expand[(x + y)^5]
```

ExpToTrig Converts exponentials to trigonometric or hyperbolic functions (not available in *Mathematica* 2.2).
```
ExpToTrig[Exp[x]]
```

Factor Factors an algebraic expression.
```
Factor[x^6 - y^6]
```

`FindMinimum` Finds a local minimum value of a function of one or more variables. To find a maximum value, apply `FindMinimum` to the negative of the function.

```
FindMinimum[x^4 - x^3 + x^2 + x - 1, {x, 0}]
FindMinimum[x^4 - 4x^3 + 6x^2 - 4x + y^2 - 4y + 5,
    {x, 0}, {y, 0}]
```

`FindRoot` Solves one or more equations numerically, using a variant of Newton's method. Works on algebraic or transcendental equations.

```
FindRoot[Exp[-x] == x^3 - 2x^2 - x + 2, {x, 1}]
```

`First` First element of a list.

```
lis = {2, 6, -3, 0}
First[lis]
```

`Flatten` Removes extra parentheses from nested lists.

```
Flatten[{2, {3, 4}, 1, {{-1, 4}, 5}}]
```

`Grad` In the `Calculus`VectorAnalysis`` package. Computes the gradient of a function.

```
Grad[x^2 + y^2 + z^2]
```

`Graphics` Creates a graphics object in 2-space

```
lyn = Line[{{0, 1}, {2, 3}}]; Show[Graphics[lyn]]
```

`Graphics3D` Creates three-dimensional graphics objects.

```
cube = Cuboid[{0, 0, 0}, {1, 2, 3}]
Show[Graphics3D[cube]]
```

`GraphicsArray` Assembles a list of graphics objects into an array.

```
Show[GraphicsArray[{Graphics[Disk[{0, 0}, 1]],
    Graphics[Circle[{0, 0}, 1]]}]];
```

`If` Used for if/then/else conditional statements.

```
f[x_] = If[x > 0, x^2, -x^2]; Plot[f[x], {x, -1, 1}]
```

`ImplicitPlot` In the `Graphics`ImplicitPlot`` package. Plots curves that are given implicitly.

```
ImplicitPlot[x^2 - y^2 == 1, {x, -3, 3}, {y, -3, 3}]
```

`Integrate` Integrates an expression symbolically. The symbol ∫ from the BasicInput palette has the same effect. Indefinite, definite, improper, and multiple integrals are all possible.

```
Integrate[x^3*Cos[x + 1], x]
Integrate[Sin[x]Exp[-x], {x, 0, Pi}]
Integrate[Exp[-x^2/2], {x, 0, Infinity}]
Integrate[x^2 - Sin[y], {x, 0, 1}, {y, x, Pi}]
```

JacobianDeterminant In the `Calculus'VectorAnalysis'` package. The Jacobian factor for integration in a specified coordinate system.

```
JacobianDeterminant[Spherical[rho, phi, theta]]
```

Join Concatenates lists together.

```
biglis = {1, 3, -7, 9, 0}; littlis = {2, 4};
Join[biglis, littlis]
```

Laplacian In the `Calculus'VectorAnalysis'` package. Computes the Laplacian of a scalar function.

```
Laplacian[(x^2 + y^2 + z^2)^2]
```

Last Last element of a list.

```
lis2 = {2, 4, 6, 8}
Last[lis2]
```

Length Length of a list.

```
rootsof1 = Solve[z^10 == 1];
Length[rootsof1]
```

Limit Evaluates the limit of an expression.

```
Limit[Sin[x]/x, x -> 0]
```

Line Draws a line with the indicated endpoints.

```
Show[Graphics[Line[{{1, 2}, {3, 4}}]]]
```

ListPlot Plots individual points in 2-space.

```
ListPlot[{{1, 3}, {2, -1}, {3, 5}},
    PlotStyle -> {PointSize[0.03]}]
```

LogLogPlot In the `Graphics'Graphics'` package. Draws a log-log plot.

```
LogLogPlot[x^3, {x, 1, 4}]
```

Max Finds the maximum element in a list.

```
lis3 = {4, -2, 0, 5.23}; Max[lis3]
```

Min Finds the minimum element in a list.

```
lis3 = {4, -2, 0, 5.23}; Min[lis3]
```

Module Programming construct; used to write programs with local variables.

```
filledcurve[f_, a_, b_] :=
    Module[{curve, x},
    curve = Plot[f[x], {x, a, b}, DisplayFunction -> Identity];
    Show[curve, Table[Graphics[Line[{{x, 0}, {x, f[x]}}]],
    {x, a, b, (b - a)/20}],
    DisplayFunction -> $DisplayFunction]]
filledcurve[Sin, 0, Pi]
```

N Evaluates an expression numerically. You can specify the desired accuracy.

```
N[Exp[-Pi^2]]
N[Sin[1], 20]
```

NDSolve Numerical differential equation solver.

```
NDSolve[{y'[x] == y[x]^3*Sin[x], y[0] == 2},
    y[x], {x, 0, 0.5}]
NDSolve[{x'[t] == -y[t]^2, y'[t] == x[t] + y[t],
    x[0] == 1, y[0] == 0}, {x[t], y[t]}, {t, 0, 1}]
```

NIntegrate Numerically approximates a definite integral (using something akin to Simpson's rule, but more complicated).

```
NIntegrate[Exp[-x^2]/x^2, {x, 1, 2}]
NIntegrate[Exp[-(x^2 + y^2)], {x, 0, 1},
    {y, 0, Sqrt[1 - x^2]}]
```

NSolve Finds numerical solutions of one or more polynomial equations.

```
NSolve[x^5 - x^2 + 1 == 0, x]
```

ParametricPlot Plots a curve parametrically in 2-space.

```
ParametricPlot[{t - Sin[t], 1 - Cos[t]}, {t, 0, 6Pi}]
```

ParametricPlot3D Plots a curve or surface parametrically in 3-space.

```
ParametricPlot3D[{t*Cos[t], t*Sin[t], t}, {t, 0, 4Pi}]
ParametricPlot3D[{Cosh[u]Cos[v], Cosh[u]Sin[v], Sinh[u]},
    {u, -1, 1}, {v, 0, 2Pi}]
```

Plot Graphs a function, or a list of functions.

```
Plot[x^3 + x^2 + x + 1, {x, -2, 2}]
Plot[{x^4 - 4x^2, 4Exp[-x^2]}, {x, -3, 3}]
```

Plot3D Graphs a function of two variables.

```
Plot3D[x^2 + y^2, {x, -1, 1}, {y, -1, 1}]
```

PlotGradientField In the `Graphics'PlotField'` package. Plots the gradient field of a function of two variables.

```
PlotGradientField[x^2 + y^2, {x, -1, 1}, {y, -1, 1}]
```

PlotGradientField3D In the `Graphics'PlotField3D'` package. Plots the gradient field of a function of three variables.

```
PlotGradientField3D[x^2 + y^2 + z^2, {x, -1, 1},
    {y, -1, 1}, {z, -1, 1}]
```

Point Draws a point at the indicated coordinates. You may want to give it some thickness to make it visible.

```
pt = Point[{1, 2}];
    Show[Graphics[{AbsolutePointSize[10], pt}]]
```

PolarPlot In the `Graphics`Graphics`` package. Plots one or more functions $r = f(\theta)$ in polar coordinates.

```
PolarPlot[1 + Cos[theta], theta, 0, 2Pi]
PolarPlot[{4/(2 + Cos[t]), 4 Cos[t] - 2}, {t, 0, 2 Pi}]
```

Polygon Graphics primitive for generating a polygon.

```
poly = Polygon[{{1, 0}, {0, 1}, {1, 1}}];
Show[Graphics[poly]]
```

Random Generates a (pseudo-)random number. With no arguments, gives a real number between 0 and 1. With arguments, gives an integer or a real or complex number in a specified range.

```
Random[]
Random[Integer, {1, 10}]
```

Remove Removes a symbol completely, so that *Mathematica* no longer recognizes its name. Needed, for example, when a command located in a package is accidentally invoked before the relevant package is loaded.

```
Show[Graphics3D[Torus[]]]
Remove[Torus]
<<Graphics`Shapes`
Show[Graphics3D[Torus[]]]
```

SetCoordinates In the `Calculus`VectorAnalysis`` package. Sets the coordinates in a desired coordinate system.

```
SetCoordinates[Cylindrical[r, theta, z]]
```

Series Computes the Taylor series of a function around a specified point, out to a specified order.

```
Series[ArcTan[x], {x, 0, 10}]
Series[Sin[x*y], {x, 0, 10}, {y, 0, 10}]
```

Show Displays several graphics simultaneously.

```
a = Plot[x^2, {x, -1, 1}];
b = ParametricPlot[{Exp[-t/10]Cos[t], Exp[-t/10]Sin[t]},
    {t, 0, 10}];
Show[{a, b}]
```

Simplify Simplifies complex algebraic expressions.

```
Simplify[(x^2 - y^2)/(x - y)]
```

Solve Finds exact solutions of an algebraic equation, or a set of equations.

```
Solve[x^2 + 5x + 7 == 0, x]
Solve[{x^2 + y^2 == 1, x + y == 1/2}, {x, y}]
```

Sort Sorts a list, putting numbers in numerical order.

```
Sort[{2, 6, 9, -3, 4}]
```

SphericalPlot3D In the `Graphics`ParametricPlot3D`` package. Plots surfaces given by equations in spherical coordinates.

```
SphericalPlot3D[1, {theta, 0, Pi}, {phi, 0, 2Pi}]
```

Sum Sums a series.

```
Sum[(1/2)^j, {j, 0, 10}]
Sum[x^j/j!, {j, 0, Infinity}]
```

Table Creates a list; useful for producing vectors.

```
Table[j, {j, 1, 10}]
```

Text Used with `Show` and `Graphics` to add text to a graph at a specified location.

```
words = Text["inflection pt", {Pi, 0}];
a := Plot[Sin[x], {x, 0, 2Pi}, Axes -> None]
Show[a, Graphics[words]]
```

Together Collects terms together over a common denominator.

```
Together[1 + 1/x]
```

// Sends the output of one command through as input to another command. For example, adding `// N` to the end of a *Mathematica* command line has the effect of asking for a numerical value, and adding `// Simplify` to the end of a *Mathematica* command line has the effect of asking for a simplified expression.

```
Solve[x^2 + 5x + 7 == 0, x] // N
Integrate[(x^4 + x^2)*Sin[x], x] // Simplify
```

/. Replacement operator. Orders *Mathematica* to use the transformation rule following it to substitute into the expression preceding it.

```
xsols = Solve[x^2 - 4 == 0, x]
x^3 + 1 /. xsols[[1]]
```

Options and Directives

AbsolutePointSize Specifies point size in printer's points.

```
pt = Point[{1, 1}];
    Show[Graphics[{AbsolutePointSize[10], pt}]]
```

AbsoluteThickness Specifies line thickness in printer's points.

```
lne = Line[{{1, 1}, {2,2}}];
    Show[Graphics[{AbsoluteThickness[10], lne}]]
```

AspectRatio Height-to-width ratio of a plot. The default is `1/GoldenRatio`, about .6. `AspectRatio -> 1` renders the height and width of a plot the same.

```
ParametricPlot[{2Cos[t], Sin[t]}, {t, 0 , 2Pi},
    AspectRatio -> 1]
```

Automatic When used as a value for `AspectRatio`, sets the scales on the x- and y-axes to be the same.

```
Plot[Sin[x], {x, 0, 3Pi}, AspectRatio -> Automatic]
```

`Axes` Specifies whether axes should be drawn in a plot.

```
Plot[Log[x], {x, 1, 10}, Axes -> False]
```

`AxesEdge` In three-dimensional graphics, specifies on what edges of the bounding box the axes should be drawn.

```
ParametricPlot3D[{Cosh[t], Sinh[t], t}, {t, -1, 1},
    AxesEdge -> {{1, 1}, {1, 1}, {1, 1}}]
```

`AxesLabel` Labels the coordinate axes in a plot.

```
Plot[Cosh[x], {x, -2, 2}, AxesLabel -> {"x", "y"}]
```

`AxesOrigin` Specifies where the axes should intersect.

```
Plot[Tan[x], {x, 0, Pi/4}, AxesOrigin -> {Pi/8, 0}]
```

`Boxed` Specifies whether a three-dimensional graphic is encased in a box. Default is `True`.

```
ParametricPlot3D[{Cosh[t], Sinh[t], t}, {t, -1, 1},
    Boxed -> False]
```

`ClipFill` Specifies whether or not `Plot3D` should fill in (with flat regions) the parts of the graph where the function being plotted is out of the `PlotRange`.

```
Plot3D[1/(x^2 + y^2), {x, -1, 1}, {y, -1, 1},
    ClipFill -> None]
```

`ColorFunction` Specifies coloring scheme.

```
ContourPlot[x^2 + y^2, {x, -1, 1}, {y, -1, 1},
    ColorFunction -> Hue]
```

`ContourShading` Specifies whether to shade regions between contours.

```
ContourPlot[x^2 + y^2, {x, -1, 1}, {y, -1, 1},
    ContourShading -> False]
```

`ContourStyle` Specifies a plotting style in `ContourPlot`.

```
ContourPlot[x^2 + y^2, {x, -1, 1}, {y, -1, 1},
    ContourShading->False, ContourStyle->Dashing[{0.1}]]
```

`Contours` Specifies which (or how many) contours to show in `ContourPlot`.

```
ContourPlot[x*y, {x, 0, 1}, {y, 0, 1}, Contours -> {0.5}]
ContourPlot[x*y, {x, 0, 1}, {y, 0, 1}, Contours -> 40]
```

`Dashing` Directive for dashed lines.

```
Plot[Sin[x], {x, 0, 2Pi}, PlotStyle -> Dashing[{0.1}]]
```

`Direction` Option for `Limit`, making it possible to compute one-sided limits. Use `Direction` -> 1 to compute the limit from the left, `Direction` -> -1 to compute the limit from the right.

```
Limit[1/x, x -> 0, Direction -> 1]
```

`DisplayFunction` Controls whether a graphics command displays its output; useful in suppressing extra output when using the `Show` command.

```
a:= Plot[Sin[x], {x, 0, 2Pi}, DisplayFunction -> Identity]
b:= Plot[Cos[x], {x, 0, 2Pi}, DisplayFunction -> Identity]
Show[{a, b}, DisplayFunction -> $DisplayFunction]
```

`GrayLevel` Determines the shade of gray in which to plot a curve, with 0 representing black and 1 representing white.

```
Plot[{x, Log[x]}, {x, 1, 10},
    PlotStyle -> {GrayLevel[0.2], GrayLevel[0.8]}]
```

`Hue` Specifies a color, on a scale from 0 to 1.

```
Plot[Sin[x], {x, 0, 2Pi}, PlotStyle->{Hue[Random[]]}]
```

`MaxIterations` Controls the number of iterations *Mathematica* will allow when running a numerical routine (like `FindRoot`).

```
FindRoot[Exp[0.04(x - 2000)] == 100, {x, 1},
    MaxIterations -> 30]
```

`MaxSteps` Specifies the maximum number of steps in `NDSolve`. Default is 1000.

```
NDSolve[{y''[x] - y'[x] + x*y[x] == 0, y[0] == 0,
    y'[0] == 1}, y[x], {x, 0, 30}, MaxSteps -> 1500]
```

`Mesh` Specifies whether a mesh is drawn in a surface graph.

```
Plot3D[x*y, {x, -1, 1}, {y, -1, 1}, Mesh -> False]
```

`Method` Selects a numerical algorithm in `FindMinimum` or `NIntegrate` (not available in *Mathematica* 2.2).

```
FindMinimum[x^20 + Exp[-100x^2], {x, 0.1},
    Method -> Newton]
NIntegrate[Exp[-Sqrt[x]]*Sin[x], {x, 0, Infinity},
    Method -> Oscillatory]
```

`PlotJoined` Connects the dots in a `ListPlot`.

```
ListPlot[{2, 1, 3, -2, 4}, PlotJoined -> True]
```

`PlotLabel` Labels a plot.

```
Plot[x^2, {x, -2, 2}, PlotLabel -> "Parabola"]
```

`PlotPoints` The number of points where an expression is sampled in plotting. The default value for `Plot` is 25. For `ContourPlot` it is 15.

```
Plot[Sin[10x^2], {x, 0, Pi}, PlotPoints -> 50]
```

`PlotRange` Specifies the range of values of the dependent variable in `Plot` or `Plot3D`. Can be used to specify the ranges of all the variables in `ParametricPlot` or `ParametricPlot3D`.

```
Plot[Sec[x], {x, 0, Pi/2}, PlotRange -> {0, 10}]
Plot[x^4 - 3x^2 + 2, {x, -5 , 5}, PlotRange -> All]
```

```
Plot3D[1/(x^2 + y^2), {x, -1, 1}, {y, -1, 1},
    PlotRange -> {0, 2}]
ParametricPlot[{t*Cos[t], t*Sin[t]}, {t, 0 , 100},
    PlotRange -> {{-2, 2}, {-3, 3}}]
```

PlotStyle Specifies a style in a two-dimensional plot command.
```
Plot[{Exp[-x^2], x^4}, {x, -2, 2}, PlotStyle ->
    {GrayLevel[0.2], GrayLevel[0.8]}]
ParametricPlot[{Cosh[t], Sinh[t]}, {t, -1, 1},
    PlotStyle -> Thickness[0.02]]
```

PointSize Specifies (relative) point size.
```
ListPlot[{{0,1}, {1,3}, {2, -1}},
    PlotStyle -> PointSize[0.05]]
```

RGBColor Specifies a color. `RGBColor[1,0,0]` means red; `RGBColor[0,1, 0]` means green; `RGBColor[0,0,1]` means blue.
```
Plot[{x, x^2, x^3}, {x, 0,1}, PlotStyle ->
    {RGBColor[1, 0, 0],RGBColor[0, 1, 0],RGBColor[0, 0, 1]}]
```

ScaleFunction Rescales arows in `PlotVectorField`, `PlotGradientField` and `PlotGradientField3D`. Its argument must be a pure function. Use a constant for equal-length arrows. Keep your ranges close to square, or utilize `AspectRatio` -> 1.
```
PlotVectorField[{y, -x}, {x, 0, 1}, {y, 0, 1},
    ScaleFunction -> (1&)]
```

Thickness Specifies (relative) line thickness.
```
ListPlot[{{0, 1}, {1, 3}, {2, -1}}, PlotJoined -> True,
    PlotStyle -> Thickness[0.05]]
```

Ticks Controls the placement of tick marks along axes in a graph.
```
Plot[Sqrt[x^2 + 1], {x, -2, 2},
    Ticks -> {{-1, 0, 1}, Automatic}]
```

VectorHeads Option in `Graphics`PlotField3D`, specifying whether or not vectors should be drawn with arrowheads. If so, `PlotPoints` should be kept small.
```
PlotVectorField3D[{x , y, z}, {x, 0, 2}, {y, 0, 2},
    {z, 0, 2}, PlotPoints-> 5, VectorHeads -> True]
```

ViewPoint Specifies a point from which to view three-dimensional graphics.
```
Plot3D[-x^2 + y^2, {x, -1, 1}, {y, -1, 1},
    ViewPoint -> {1, 1, 1}]
```

WorkingPrecision Specifies the precision in a numerical algorithm (such as employed in `FindMinimum`). If round-off error seems to be a problem, try increasing this option.

```
FindMinimum[x^4 - 4x^3 + 6x^2 - 4x + 5 + y^2 - 4y, {x, 0},
    {y, 0}, WorkingPrecision -> 25]
```

Built-in Functions

Abs Absolute value function.

ArcCos Inverse cosine function.

ArcCosh Inverse hyperbolic cosine function.

ArcSin Inverse sine function.

ArcSinh Inverse hyperbolic sine function.

ArcTan Inverse tangent function. With two arguments, computes the angular variable in polar coordinates.

```
ArcTan[-1, -1]
```

ArcTanh Inverse hyperbolic tangent function.

Cos Cosine function.

Cosh Hyperbolic cosine function.

Cot Cotangent function.

Csc Cosecant function.

Exp The exponential function. The expressions **Exp[x]** and **E^x** have exactly the same meaning.

Factorial The factorial of a number. Can also be denoted by an exclamation point !.

```
5!
```

Im Takes the imaginary part of a complex number.

Log The *natural* logarithm, or logarithm to the base *e*. With two arguments, computes the logarithm to another base.

```
Log[10, 10000]
```

Re Takes the real part of a complex number.

Sec Secant function.

Sech Hyperbolic secant function.

Sign The sign of a number.

Sin Sine function.

Sinh Hyperbolic sine function.

Sqrt Square root function. The symbol $\sqrt{}$ from the BasicInput palette has the same effect.

Tan Tangent function.

`Tanh` Hyperbolic tangent function.

Built-in Constants

`E` The number $e = 2.718281828\ldots$.

`I` The imaginary unit $i = \sqrt{-1}$. Even in problems that don't appear to involve complex numbers, *Mathematica* sometimes will find complex roots to equations or will express integrals in terms of complex-valued functions.

`Infinity` Self-explanatory. *Mathematica* recognizes this as a valid limit in definite integrals, sums, etc. The symbol ∞ from the standard palette has the same meaning.

`Pi` The number π. The symbol π from the BasicInput palette has the same meaning.

Packages

`Calculus`VectorAnalysis`` Needed for the commands `Grad`, `Div`, `Curl`, `Laplacian`, and `CrossProduct`.

`Graphics`Arrow`` Needed for `Arrow`.

`Graphics`ContourPlot3D`` Needed for `ContourPlot3D`.

`Graphics`Graphics`` Needed for `LogLogPlot` and `PolarPlot`.

`Graphics`ImplicitPlot`` Needed for `ImplicitPlot`.

`Graphics`ParametricPlot3D`` Needed for `CylindricalPlot3D`, for an enhanced version of `ParametricPlot3D`, and for `SphericalPlot3D`.

`Graphics`PlotField`` Needed for `PlotGradientField`.

`Graphics`PlotField3D`` Needed for `PlotGradientField3D`.

`Graphics`Shapes`` Extra routines that are handy for displaying spheres, cones, tori, etc.

`Graphics`SurfaceOfRevolution`` Extra routines that are handy for displaying surfaces of revolution.

`Miscellaneous`RealOnly`` Filters out nonreal solutions to equations.

Sample Notebook Solutions

■ Problem Set A, Problem 3

■ (a)

```
Clear[f, x]

f[x_] = x * Sin[x]

x Sin[x]

Plot[{f[x], 1}, {x, 0, 15}];
```

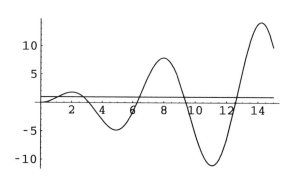

The solutions we are looking for are located roughly at $x = 1, 3, 6.5, 9.5$, and 12.5.

▪ (b)

We are trying to solve $\sin x = 1/x$. But for large $|x|$, $1/x$ is close to 0, so the solution is close to the solution of $\sin x = 0$. Now the sine function vanishes at every integral multiple of π. So the solutions are roughly at $n\,\pi$, with n a positive integer.

▪ (c)

We apply **FindRoot** with each of the approximate values of x found above used as a starting value in turn.

```
startpts = {1, 3, 6.5, 9.5, 12.5}

{1, 3, 6.5, 9.5, 12.5}

actualroots =
  Table[x /. FindRoot[f[x] == 1, {x, startpts[[j]]}],
    {j, 1, 5}]

{1.11416, 2.7726, 6.43912, 9.31724, 12.6455}
```

The positions of the roots get closer and closer to multiples of π.

```
Table[actualroots[[j]] - (j - 1) * Pi, {j, 2, 5}]

{-0.368988, 0.155932, -0.107535, 0.079162}
```

▪ Problem Set B, Problem 9

```
Clear[F, G, a, b, c, d]

norm[x_] := Sqrt[x.x]

unitVector[x_] := x / norm[x]
```

```
F[a_, b_] := unitVector[a - b]

G[a_, b_] := norm[a - b]
```

The function *F* gives a unit vector in the direction from *b* to *a*. The function *G* measures the length of the vector connecting *b* to *a*.

```
a := {1, 1, 1}; b := {2, 3, 3}; c := {4, 2, 5};
d := {3, 0, 3};

F[a, b] . F[c, d]

    -1
```

This shows the unit vector $(a - b)/\|(a - b)\|$ is parallel to the unit vector $(c - d)/\|(c - d)\|$. Similarly

```
F[a, d] . F[b, c]

    1
```

So the quadrilateral determined by the four points is a parallelogram.

```
G[a, b]

    3

G[b, c]

    3
```

So adjacent sides have the same length; the parallelogram is a rhombus.

```
(a - b) . (b - c)

    4
```

But adjacent sides are not perpendicular; the rhombus is not a square.

■ Problem Set C, Problem 1, Parts (a), (d), and (f)

```
Clear[t, r]

<< Calculus`VectorAnalysis`

vectorLength[x_] := Sqrt[Simplify[x . x]]
unitVector[x_] := Simplify[x / vectorLength[x]]

velocity[r_] := Simplify[D[r, t]]
acceleration[r_] := Simplify[D[r, {t, 2}]]
speed[r_] := vectorLength[velocity[r]]
UT[r_] := unitVector[velocity[r]]
UN[r_] := unitVector[D[UT[r], t]]
UB[r_] := Simplify[CrossProduct[UT[r], UN[r]]]

curvature[r_] :=
 Simplify[vectorLength[D[UT[r], t]] / speed[r]]
torsion[r_] := tau /. Simplify[First[Solve[
      D[UB[r], t] == -tau * speed[r] * UN[r], tau]]]
```

■ (a) Cardioid

```
cardioid = {2Cos[t]*(1 + Cos[t]), 2Sin[t]*(1 + Cos[
t])}
```

$\{2 \, Cos[t] \, (1 + Cos[t]), \, 2 \, (1 + Cos[t]) \, Sin[t]\}$

```
speed[cardioid]
```

$2 \sqrt{2} \, \sqrt{1 + Cos[t]}$

```
UT[cardioid]
```

$\{-\dfrac{Sin[t] + Sin[2\,t]}{\sqrt{2} \, \sqrt{1 + Cos[t]}}, \, \dfrac{Cos[t] + Cos[2\,t]}{\sqrt{2} \, \sqrt{1 + Cos[t]}}\}$

UN[cardioid]

$$\left\{ -\frac{2\sqrt{2}\ \text{Cos}[\frac{t}{2}]^4\ (-1 + 2\,\text{Cos}[t])}{(1 + \text{Cos}[t])^{3/2}}, \right.$$

$$\left. -\frac{2\sqrt{2}\ \text{Cos}[\frac{t}{2}]^3\ \text{Sin}[\frac{3t}{2}]}{(1 + \text{Cos}[t])^{3/2}} \right\}$$

UB[Join[cardioid, {0}]]

{0, 0, 1}

curvature[cardioid]

$$\frac{3}{4\sqrt{2}\ \sqrt{1 + \text{Cos}[t]}}$$

torsion[Join[cardioid, {0}]]

0

Well, we knew that; since this is a plane curve, the torsion vanishes.

ParametricPlot[Evaluate[cardioid], {t, 0, 2Pi}]

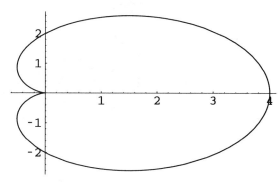

- Graphics -

The curvature blows up at $t = \pi$, which corresponds to the cusp in the picture.

▪ (d) Hyperbolic Helix

```
hyp = {Sqrt[1 + t^2/2], t/Sqrt[2], ArcSinh[t/Sqrt[
2]]}
```

$$\{\sqrt{1 + \frac{t^2}{2}}, \frac{t}{\sqrt{2}}, \text{ArcSinh}\left[\frac{t}{\sqrt{2}}\right]\}$$

speed[hyp]

1

This is a unit speed curve.

UT[hyp]

$$\{\frac{t}{\sqrt{2}\sqrt{2 + t^2}}, \frac{1}{\sqrt{2}}, \frac{1}{\sqrt{2 + t^2}}\}$$

UN[hyp]

$$\{\sqrt{2}\sqrt{\frac{1}{(2 + t^2)^2}}\sqrt{2 + t^2}, 0, -t\sqrt{\frac{1}{(2 + t^2)^2}}\sqrt{2 + t^2}\}$$

UB[hyp]

$$\{-\frac{t\sqrt{\frac{1}{(2+t^2)^2}}\sqrt{2 + t^2}}{\sqrt{2}}, \frac{\sqrt{\frac{1}{(2+t^2)^2}}(2 + t^2)}{\sqrt{2}},$$

$$-\sqrt{\frac{1}{(2 + t^2)^2}}\sqrt{2 + t^2}\}$$

curvature[hyp]

$$\sqrt{\frac{1}{(2 + t^2)^2}}$$

torsion[hyp]

$$\frac{1}{2+t^2}$$

This curve is a helix since the ratio τ/κ is the constant 1, as we see from

(curvature[hyp] / torsion[hyp]) ^ 2

1

ParametricPlot3D[Evaluate[hyp], {t, -5, 5}]

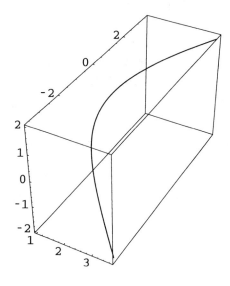

- Graphics3D -

If we set $t = \sqrt{2} \sinh u$, then we can solve for the coordinate functions in terms of t:

hyp /. t -> Sqrt[2] * Sinh[u] // Simplify

$\left\{ \sqrt{\text{Cosh[u]}^2}, \text{Sinh[u]}, \text{ArcSinh[Sinh[u]]} \right\}$

So clearly the new parametrization is: (cosh u, sinh u, u) and in the same way that (cos t, sin t, t) represents a circular helix, the current parametrization represents a "hyperbolic helix".

▪ (f) Viviani's Curve

```
viviani = {1 + Cos[t], Sin[t], 2 Sin[t / 2]}
```

$$\left\{1 + \text{Cos}[t], \text{Sin}[t], 2 \text{Sin}\left[\frac{t}{2}\right]\right\}$$

```
ParametricPlot3D[
 Evaluate[{viviani, {1 + Cos[t], Sin[t], 0},
    {2 Cos[t / 2], 0, 2 Sin[t / 2]},
    {0, 2 Cos[t / 2], 2 Sin[t / 2]},
    {2 Cos[t / 2], 2 Sin[t / 2], 0}}],
 {t, 0, Pi}, Boxed -> False, Ticks -> None,
 AxesEdge -> {{-1, -1}, {-1, -1}, {-1, -1}},
 PlotRange -> {{0, 2}, {0, 2}, {0, 2}},
 ViewPoint -> {10, 3, 1}]
```

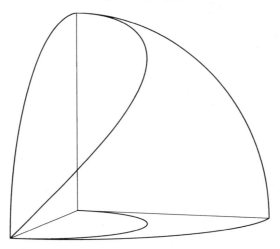

```
- Graphics3D -
```

```
speed[viviani]
```

$$\frac{\sqrt{3 + \text{Cos}[t]}}{\sqrt{2}}$$

Since the speed never vanishes, this is a smooth curve.

UT[viviani]

$$\left\{-\frac{\sqrt{2}\,\text{Sin}[t]}{\sqrt{3+\text{Cos}[t]}},\ \frac{\sqrt{2}\,\text{Cos}[t]}{\sqrt{3+\text{Cos}[t]}},\ \frac{\sqrt{2}\,\text{Cos}[\frac{t}{2}]}{\sqrt{3+\text{Cos}[t]}}\right\}$$

UN[viviani]

$$\left\{-\frac{3+12\,\text{Cos}[t]+\text{Cos}[2\,t]}{2\,(3+\text{Cos}[t])^{3/2}\,\sqrt{\frac{13+3\,\text{Cos}[t]}{(3+\text{Cos}[t])^2}}},\right.$$

$$-\frac{12\,\text{Sin}[t]+\text{Sin}[2\,t]}{2\,(3+\text{Cos}[t])^{3/2}\,\sqrt{\frac{13+3\,\text{Cos}[t]}{(3+\text{Cos}[t])^2}}},$$

$$\left.-\frac{2\,\text{Sin}[\frac{t}{2}]}{(3+\text{Cos}[t])^{3/2}\,\sqrt{\frac{13+3\,\text{Cos}[t]}{(3+\text{Cos}[t])^2}}}\right\}$$

UB[viviani]

$$\left\{\frac{\sqrt{\frac{13+3\,\text{Cos}[t]}{(3+\text{Cos}[t])^2}}\,(16\,\text{Sin}[\frac{t}{2}]+9\,\text{Sin}[\frac{3\,t}{2}]+\text{Sin}[\frac{5\,t}{2}])}{\sqrt{2}\,(26+6\,\text{Cos}[t])},\right.$$

$$\left.-\frac{4\,\text{Cos}[\frac{t}{2}]^3}{(3+\text{Cos}[t])\,\sqrt{\frac{26+6\,\text{Cos}[t]}{(3+\text{Cos}[t])^2}}},\ \frac{4}{(3+\text{Cos}[t])\,\sqrt{\frac{26+6\,\text{Cos}[t]}{(3+\text{Cos}[t])^2}}}\right\}$$

vk = curvature[viviani]

$$\frac{\sqrt{\frac{13+3\,\text{Cos}[t]}{(3+\text{Cos}[t])^2}}}{\sqrt{3+\text{Cos}[t]}}$$

vk = Sqrt[Simplify[vk^2]]

$$\sqrt{\frac{13+3\,\text{Cos}[t]}{(3+\text{Cos}[t])^3}}$$

torsion[viviani]

$$\frac{6\,\text{Cos}[\frac{t}{2}]}{13+3\,\text{Cos}[t]}$$

The curvature and torsion never blow up, reflecting the smoothness of the curve.

■ Problem Set D, Problem 3

```
Clear[r, v, a, t, w]
```

■ (a)

If the planet moves in a circular orbit, then since the gravitational force points toward the center of the circle and the velocity is tangent to the circle, the two are perpendicular to one another. By equation (1) in Chapter 4, *Kinematics*, this means the kinetic energy and the speed are constant.

■ (b)

We assume that the motion is circular at constant speed and differentiate. First we define the position vector as a function of *t*.

```
posit[t_] = {r*Cos[w*t], r*Sin[w*t]}
```
$\{r \, \text{Cos}[t \, w], \, r \, \text{Sin}[t \, w]\}$

```
a[t_] = D[posit[t], {t, 2}]
```
$\{-r \, w^2 \, \text{Cos}[t \, w], \, -r \, w^2 \, \text{Sin}[t \, w]\}$

```
sol = Solve[a[t] == -posit[t]/r^3, w]
```

Solve::ifun : Inverse functions are being used by Solve, so
    some solutions may not be found.

$$\left\{ \left\{ w \to -\frac{1}{r^{3/2}} \right\}, \, \left\{ w \to \frac{1}{r^{3/2}} \right\} \right\}$$

We see that the angular velocity and radius are related by $\omega = 1/r^{3/2}$. (Obviously we can ignore the negative solution.) Then the speed is related to the radius by:

```
v = r*w /. sol[[2]]
```

$$\frac{1}{\sqrt{r}}$$

So speed is inversely proportional to the square root of the radius of the orbit.
The period of the orbit is the time it takes to go all the way around, or

```
per = 2Pi*r/v
```

$$2 \pi r^{3/2}$$

▪ (c)

```
equations[yp0_] =
 {x''[t] == -x[t] / (x[t]^2 + y[t]^2)^(3/2),
  y''[t] == -y[t] / (x[t]^2 + y[t]^2)^(3/2),
  x[0] == 1, y[0] == 0, x'[0] == 0, y'[0] == yp0}
```

$$\left\{ x''[t] == -\frac{x[t]}{\left(x[t]^2 + y[t]^2\right)^{3/2}}, \right.$$

$$y''[t] == -\frac{y[t]}{\left(x[t]^2 + y[t]^2\right)^{3/2}}, \; x[0] == 1, \; y[0] == 0,$$

$$\left. x'[0] == 0, \; y'[0] == yp0 \right\}$$

▪ (i)

```
soli = NDSolve[equations[1], {x[t], y[t]}, {t, 0,
10}]
```

```
{{x[t] → InterpolatingFunction[{{0., 10.}}, <>][t],
  y[t] → InterpolatingFunction[{{0., 10.}}, <>][t]}}
```

```
ParametricPlot[Evaluate[{x[t], y[t]} /. soli], {t, 0,
10}, AspectRatio -> Automatic];
```

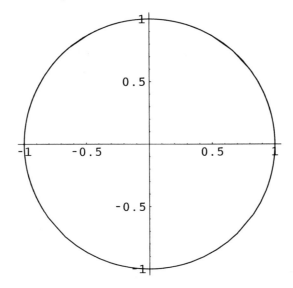

The orbit in this case is a circle.

■ **(ii)**

```
solii = NDSolve[equations[1.2], {x[t], y[t]}, {t, 0,
15}]
```

```
{{x[t] → InterpolatingFunction[{{0., 15.}}, <>][t],
   y[t] → InterpolatingFunction[{{0., 15.}}, <>][t]}}
```

```
ParametricPlot[Evaluate[{x[t], y[t]} /. solii],
  {t, 0, 15}, AspectRatio -> Automatic];
```

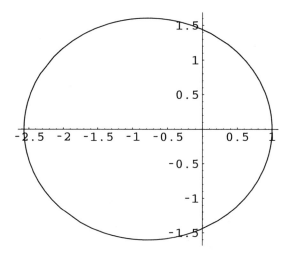

This time we get an ellipse. This is not so surprising since the object starts off tangent to the same circle as in (i), but with a little extra speed, too much to stay on the circle. We can see the ellipse more clearly if we pay attention to the tick marks. Note also we allowed more time.

■ **(iii)**

In part (iii), the initial velocity is much greater; let's see what happens.

```
soliii = NDSolve[equations[2], {x[t], y[t]},
  {t, -10, 10}]
```

```
{{x[t] → InterpolatingFunction[{{-10., 10.}}, <>][t],
    y[t] → InterpolatingFunction[{{-10., 10.}}, <>][t]}}
```

```
ParametricPlot[Evaluate[[x[t], y[t]} /. soliii],
   {t, -10, 10}];
```

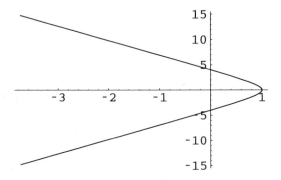

In this last case, the object follows a hyperbola. The trajectory is very close to an asymptote that is a straight line with slope around -2. Note that to see the hyperbola more clearly we also went in the direction of negative time.

■ Problem Set E, Problem 2, Parts (a) and (b)(i, ii, iv)

■ (a)

If $u_x = v_y = a$ and $u_y = -v_x = b$, then grad $u = (a, b)$ and grad $v = (-b, a)$. Hence the dot product of these is:

```
Clear[a, b, u, v, x, y, z]

{a, b} . {-b, a}

0
```

This proves that the gradients are orthogonal. Furthermore,

$$u_{xx} + u_{yy} = (u_x)_x + (u_y)_y = (v_y)_x + (-v_x)_y = v_{yx} - v_{xy} = 0,$$

by equality of the mixed second partials. A similar argument works with v.

▪ (b)

```
<< Calculus`VectorAnalysis`

SetCoordinates[Cartesian[x, y, z]]

Cartesian[x, y, z]
```

To check that functions *u* and *v* are harmonic conjugates, we subtract the second entry of the gradient of *v* from the first entry of the gradient of *u*, and add the second entry of the gradient of *u* to the first entry of the gradient of *v*. If we get 0 both times, the conditions are satisfied.

```
check[u_, v_] := {Grad[u][[1]] - Grad[v][[2]],
   Grad[v][[1]] + Grad[u][[2]]}
```

▪ (i)

```
u = x; v = y;

check[u, v]

{0, 0}

uplot = ContourPlot[u, {x, -2, 2}, {y, -2, 2},
   ContourStyle -> GrayLevel[0],
   ContourShading -> False,
   DisplayFunction -> Identity];

vplot = ContourPlot[v, {x, -2, 2}, {y, -2, 2},
   ContourStyle -> GrayLevel[0.7],
   ContourShading -> False,
   DisplayFunction -> Identity];
```

```
Show[uplot, vplot,
  DisplayFunction -> $DisplayFunction];
```

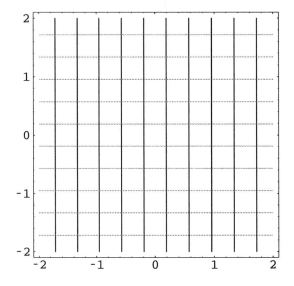

A checkerboard!

■ **(ii)**

```
u = x^2 - y^2; v = 2 x y;
```

```
check[u, v]
```

```
{0, 0}
```

```
uplot = ContourPlot[u, {x, -2, 2}, {y, -2, 2},
  ContourStyle -> GrayLevel[0],
  ContourShading -> False,
  DisplayFunction -> Identity];
```

```
vplot = ContourPlot[v, {x, -2, 2}, {y, -2, 2},
  ContourStyle -> GrayLevel[0.7],
  ContourShading -> False,
  DisplayFunction -> Identity];
```

```
Show[uplot, vplot,
  DisplayFunction -> $DisplayFunction];
```

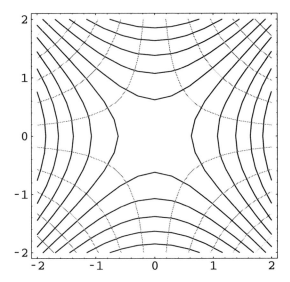

A spiderweb!

■ **(iv)**

```
u = x (1 + 1 / (x^2 + y^2)); v = y (1 - 1 / (x^2 + y^2));

check[u, v] // Simplify
```

```
{0, 0}
```

We need to increase **Contours** and **PlotPoints** to get a clear picture.

```
uplot = ContourPlot[u, {x, -2, 2}, {y, -2, 2},
  ContourStyle -> GrayLevel[0],
  ContourShading -> False,
  DisplayFunction -> Identity, Contours -> 40,
  PlotPoints -> 40];
```

```
vplot = ContourPlot[v, {x, -2, 2}, {y, -2, 2},
  ContourStyle -> GrayLevel[0.7],
  ContourShading -> False,
  DisplayFunction -> Identity, Contours -> 40,
  PlotPoints -> 40];
```

```
Show[uplot, vplot,
 DisplayFunction -> $DisplayFunction];
```

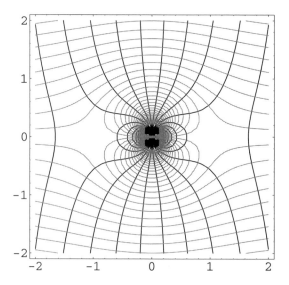

Pretty neat!

■ Problem Set F, Problem 3

```
Clear[u, v, e, f, g]
```

```
<<Calculus`VectorAnalysis`
```

```
vecLength[vec_] := Sqrt[vec . vec]
```

```
normalVector[surf_, u_, v_] :=
  CrossProduct[D[surf, u], D[surf, v]] // Simplify
```

```
unitNorm[surf_, u_, v_] := normalVector[surf, u, v] /
    vecLength[normalVector[surf, u, v]] //
  Simplify

efgInvariants[surf_, u_, v_] :=
 {ee -> D[surf, u] . D[surf, u],
   ff -> D[surf, u] . D[surf, v],
  gg -> D[surf, v] . D[surf, v],
   e -> unitNorm[surf, u, v] . D[surf, {u, 2}],
   f -> unitNorm[surf, u, v] . D[surf, u, v],
   g -> unitNorm[surf, u, v] . D[surf, {v, 2}]} //
  Simplify

shapeOperator[surf_, u_, v_] :=
 ( (ee*gg - ff^2) ^ (-1) * {{e*gg - f*ff, f*ee - e*ff},
  {f*gg - g*ff, g*ee - f*ff}} /.
  efgInvariants[surf, u, v]) // Simplify

gaussCurv[surf_, u_, v_] :=
 Det[shapeOperator[surf, u, v]] // Simplify
```

▪ (a) Ellipsoid

```
ellip = {Cos[u]Sin[v], Sin[u]Sin[v], 2Cos[v]}

{Cos[u] Sin[v], Sin[u] Sin[v], 2 Cos[v]}

ellipcurv[u_, v_] = gaussCurv[ellip, u, v]
```

$$\frac{16}{(5 - 3 \cos[2 v])^2}$$

```
ParametricPlot3D[Evaluate[ellip], {u, 0, 2Pi},
    {v, 0, Pi}, ViewPoint -> {2, 2, 2}];
```

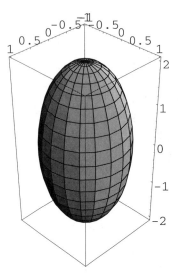

The Gaussian curvature, whose denominator is a square, is always positive. Thus the ellipse is a surface of positive curvature. At every point, in every direction, the surface bends away from the outer normal. Also, the curvature only depends upon v, because the ellipsoid is unchanged by rotation about the z-axis. The curvature is the biggest when $v = 0$ (the north pole), decreases until $v = \pi/2$ (the equator), and increases back to its largest value at the south pole when $v = \pi$. This happens because the vertical slices, which are elliptic, bend most at the poles and least at the equator.

■ (b) Elliptic Hyperboloid of One Sheet

```
ellhyp = {Cos[u]Cosh[v], Sin[u]Cosh[v]/Sqrt[2],
    Sinh[v]}
```

$$\left\{\text{Cos[u] Cosh[v]}, \frac{\text{Cosh[v] Sin[u]}}{\sqrt{2}}, \text{Sinh[v]}\right\}$$

```
ellhypcurv[u_, v_] = gaussCurv[ellhyp, u, v]
```

$$- \frac{32}{((-5 + \text{Cos}[2\,u])\,\text{Cosh}[2\,v] - 2\,\text{Sin}[u]^2)^2}$$

```
ParametricPlot3D[Evaluate[ellhyp], {u, 0, 2Pi},
    {v, -1, 1}, ViewPoint -> {2, 2, 2}];
```

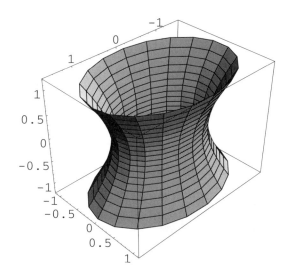

The Gaussian curvature is always negative. Thus, the hyperboloid of one sheet is a surface of negative curvature. At every point, the surface has the nature of a saddle; it bends away from the normal in one direction, but bends toward it in the perpendicular direction. Also, the curvature is an even function of v (for fixed u), because the hyperboloid is unchanged under reflection across the x-y plane. As $v \rightarrow \infty$, the Cosh[$2v$] term in the denominator $\rightarrow \infty$, so the curvature approaches zero, reflecting the increasingly flat nature of the hyperbolic profile curves in v. For $v = 0$, the curvature is

```
ellhypcurv[u, 0] // Simplify
```

$$- \frac{8}{(-3 + \text{Cos}[2\,u])^2}$$

which reflects the nature of the elliptical slices parallel to the *x-y* plane. The curvature is the most negative when $v = 0$ and $u = 0$ or π, since the ellipse corresponding to $z = 0$ has the greatest curvature on the *x*-axis.

■ (c) Hyperboloid of Two Sheets

```
hyp2 = {Cos[u]Sinh[v], Sin[u]Sinh[v], 2Cosh[v]}
```

{Cos[u] Sinh[v], Sin[u] Sinh[v], 2 Cosh[v]}

```
hyp2curv[u_, v_] = gaussCurv[hyp2, u, v]
```

$$\frac{16}{(3 - 5 \, \text{Cosh}[2 \, v])^2}$$

```
ParametricPlot3D[Evaluate[hyp2], {u, 0, 2Pi},
    {v, -1, 1}, ViewPoint -> {2, 2, 0.5}];
```

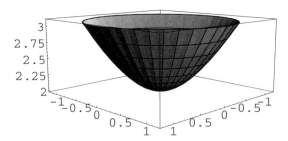

The Gaussian curvature is always positive. Thus the hyperboloid of two sheets is a surface of positive curvature. At every point, in every direction, the surface bends away from the outer normal. Also, the curvature only depends upon *v*, because the hyperboloid is unchanged by rotation about the *z*-axis. The curvature is the biggest when $v = 0$ and tends to zero as $v \to \infty$. This reflects the fact that the hyperbolas that form the vertical cross sections have the greatest curvature at their "midpoints," and tend toward a straight line (the asymptote) at infinity.

■ Problem Set G, Problem 1

```
Clear[f, x, y]
```

```
f[x_, y_ ] = x^4 - 3x*y + 2y^2
```

$x^4 - 3 x y + 2 y^2$

- **(a)**

    ```
    fx[x_, y_] = D[f[x, y], x]
    ```

    $4 x^3 - 3 y$

    ```
    fy[x_, y_] = D[f[x, y], y]
    ```

    $-3 x + 4 y$

    ```
    disc[x_, y_] = D[fx[x, y], x]* D[fy[x, y], y] -
      D[fx[x, y], y]^2
    ```

    $-9 + 48 x^2$

    ```
    critpts = Solve[{fx[x, y] == 0, fy[x, y] == 0}, {x,
    y}]
    ```

    $\left\{\left\{y \to -\dfrac{9}{16}, x \to -\dfrac{3}{4}\right\}, \{y \to 0, x \to 0\}, \left\{y \to \dfrac{9}{16}, x \to \dfrac{3}{4}\right\}\right\}$

We've found three critical points. Now we compute the discriminant at each one to classify them.

    ```
    disc[x, y] /. critpts
    ```

    $\{18, -9, 18\}$

Thus $(-3/4, -9/16)$ and $(3/4, 9/16)$ have positive discriminant and are local extrema. The point $(0, 0)$ has negative discriminant and is a saddle point. To find the types of the two local extrema, we look at the sign of, say, f_{xx}

    ```
    D[fx[x, y], x] /. {critpts[[1]], critpts[[3]]}
    ```

    $\left\{\dfrac{27}{4}, \dfrac{27}{4}\right\}$

Since f_{xx} is positive at both local extrema, they are local minimum points.

■ **(b)**

```
FindMinimum[f[x, y], {x, 0.5}, {y, 0.5}]
```

$\{-0.316406, \{x \to 0.75, y \to 0.5625\}\}$

This agrees with part (a) since 0.75 = 3/4 and 0.5625 = 9/16.

```
FindMinimum[f[x, y], {x, -0.5}, {y, 0.5}]
```

$\{-0.316406, \{x \to -0.749996, y \to -0.5625\}\}$

This basically agrees with part (a) but there is some numerical error in the calculation of *x*. We can get better accuracy by increasing the working precision.

```
FindMinimum[f[x, y], {x, -0.5}, {y, 0.5},
WorkingPrecision->20]
```

$\{-0.316406250000000000, \{x \to -0.7500000000000004099,$
$y \to -0.5625000000000010202\}\}$

That's better. Now it's clear that $x = -0.75$ and $y = -0.5625$ is correct to a large number of decimal places.

■ **(c)**

```
FindMinimum[f[x, y], {x, 0}, {y, 0}]
```

```
FindMinimum::fmgz :
  Encountered a vanishing gradient. The result returned may
    not be a minimum; it may be a maximum or a saddle point.
```

$\{0., \{x \to 0., y \to 0.\}\}$

The steepest descent method stops immediately since we began at a critical point. In this case it is indeed a saddle point, as we know from (a).

■ **(d)**

The minimum value of *f* at both local minima is (approximately) −0.316406

```
Plot3D[f[x, y], {x, -1, 1}, {y, -1, 1}];
```

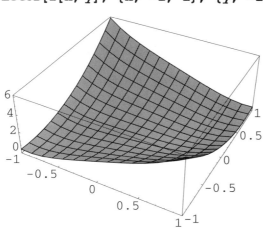

No further pictures are provided, but even with different choices of the
ViewPoint and various **PlotRange** specifications, it is still virtually
impossible to discern either of the extrema or the saddle point. On the other
hand, the symbolic calculations in (a), and the symbolic/numeric calculations
in (b), yielded very accurate information on the location of the critical points
and the values of the function there.

■ Problem Set H, Problem 2

```
Clear[x, y, r, t]

<< Graphics`Graphics`

toPolar = {x -> r * Cos[t], y -> r * Sin[t]}
{x → r Cos[t], y → r Sin[t]}

xp[r_] := r * Cos[t]
yp[r_] := r * Sin[t]
```

```
polarRegion[a_, b_, R1_, R2_] :=
 Module[{plot1}, plot1 = PolarPlot[{R1, R2},
    {t, a, b}, PlotStyle -> {{}, Thickness[0.008]},
    DisplayFunction -> Identity];
  Show[plot1, Table[Graphics[
    Line[{{xp[R1], yp[R1]}, {xp[R2], yp[R2]}}]],
    {t, a, b, (b - a) / 30}],
   DisplayFunction -> $DisplayFunction]]

verticalRegion[a_, b_, f_, g_] :=
 Module[{twocurves},
  twocurves = Plot[{f, g}, {x, a, b},
    PlotStyle -> {{}, Thickness[0.008]},
    DisplayFunction -> Identity];
  Show[twocurves,
   Table[Graphics[Line[{{x, f}, {x, g}}]]],
   {x, a, b, (b - a) / 20}],
   DisplayFunction -> $DisplayFunction]]
```

■ **(a)**

```
polarRegion[0, 2 Pi, 0, 2 (1 + Sin[t])]
```

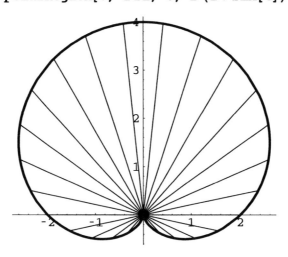

- Graphics -

```
integrand = x^2 * y^2 /. toPolar
```

$r^4 \text{Cos}[t]^2 \text{Sin}[t]^2$

```
Integrate[integrand * r, {t, 0, 2 Pi},
  {r, 0, 2 (1 + Sin[t])}]
```

$\dfrac{143\,\pi}{4}$

```
N[%]
```

112.312

■ (b)

```
polarRegion[0, Pi / 2, 0, Sin[t] ^2]
```

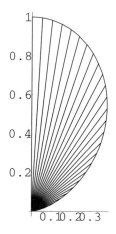

- Graphics -

```
Cos[x^2 + y^2] /. toPolar // Simplify
```

$\text{Cos}[r^2]$

```
Integrate[r * Cos[r^2], {t, 0, Pi / 2},
 {r, 0, Sin[t] ^2}]
```

$$\frac{3}{32} \, \pi \, \text{HypergeometricPFQ}\left[\left\{\frac{5}{8}, \frac{7}{8}, \frac{9}{8}, \frac{11}{8}\right\}, \left\{\frac{3}{4}, 1, \frac{5}{4}, \frac{3}{2}, \frac{3}{2}\right\}, -\frac{1}{4}\right]$$

```
N[%]
```

0.266125

■ (c)

```
verticalRegion[3 / Sqrt[2], 3, 0, Sqrt[9 - x^2]]
```

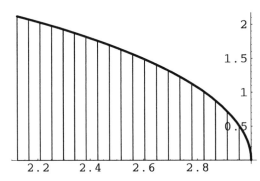

- Graphics -

It's a little hard to see that the top curve is a circle. Its equation in Cartesian coordinates is $y = \sqrt{9 - x^2}$ or $x^2 + y^2 = 9$; that is, the circle of radius 3 centered at the origin. So the region described in polar coordinates will go from the vertical line to the circle, and then between $\theta = 0$ and the angle determined by the intersection of the circle and the vertical line. We will compute these missing ingredients.

The equation of the line is $x = 3/\sqrt{2}$.

```
x == 3 / Sqrt[2] /. toPolar
```

$$r \, Cos[t] == \frac{3}{\sqrt{2}}$$

```
Solve[%, r]
```

$$\left\{\left\{r \to \frac{3 \, Sec[t]}{\sqrt{2}}\right\}\right\}$$

```
Solve[{r == 3 Sec[t] / Sqrt[2], r == 3}, t]
```

Solve::ifun : Inverse functions are being used by Solve, so
    some solutions may not be found.

$$\left\{\left\{t \to -\frac{\pi}{4}\right\}, \left\{t \to \frac{\pi}{4}\right\}\right\}$$

The angle is $\pi/4$; so

```
polarRegion[0, Pi / 4, 3 Sec[t] / Sqrt[2], 3]
```

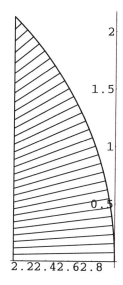

- Graphics -

Since $x^2 + y^2 = r^2$, the integrand is $r^2 \, r = r^3$.

```
Integrate[r^3, {t, 0, Pi/4},
 {r, 3 Sec[t]/Sqrt[2], 3}]
```

$$\frac{27}{16}\,(-4 + 3\,\pi)$$

```
N[%]
```

9.15431

■ (d)

Let's see if we can get by without drawing the region in Cartesian coordinates. As in (c), the equation $y = \sqrt{1 - x^2}$, or $x^2 + y^2 = 1$, represents the circle of radius 1 centered at the origin. Everything lives inside it. The inner boundary is given by

```
y == Sqrt[x - x^2] /. toPolar // Simplify
```

$$r\,\text{Sin}[t] == \sqrt{r\,\text{Cos}[t]\,(1 - r\,\text{Cos}[t])}$$

```
Solve[%, r]
```

$$\{\{r \to 0\},\ \{r \to \text{Cos}[t]\}\}$$

The second equation is the curve we want; it is a circle of radius 1/2, centered at (1/2, 0). We leave it to you to corroborate that the limits of integration are

```
polarRegion[0, Pi / 2, Cos[t], 1]
```

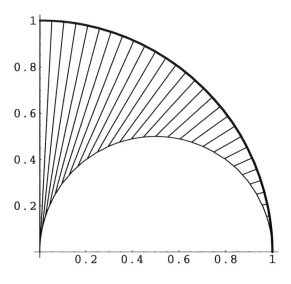

- Graphics -

Since $x^2 + y^2 = r^2$, the integrand is $(r^2)^{3/2} r = r^4$.

```
Integrate[r^4, {t, 0, Pi / 2}, {r, Cos[t], 1}]
```

$$\frac{-128 + 120\,\pi}{1200}$$

```
N[%]
```

0.207493

■ Problem Set I, Problem 1

The potential $\phi 1$ due to a charge Q at $(0, 0, 1)$ and the potential $\phi 2$ due to a charge Q at $(0, 0, -1)$ will exactly cancel each other on the *x-y* plane. So the sum ϕ will have value 0 on the *x-y* plane, which is exactly the condition we want for a conducting plate. Thus we can compute ϕ and the field exactly.

```
Clear[x, y, z, q]
```

```
phi[x_, y_, z_, q_] = -q/Sqrt[x^2 + y^2 + (z-1)^2] +
  q/Sqrt[x^2 + y^2 + (z+1)^2]
```

$$-\frac{q}{\sqrt{x^2 + y^2 + (-1 + z)^2}} + \frac{q}{\sqrt{x^2 + y^2 + (1 + z)^2}}$$

Let's check that ϕ is zero on the *x-y* plane.

```
Simplify[phi[x, y, 0, q]]
```

0

The electric field is the negative of the gradient of ϕ. Let's use *Mathematica*'s vector calculus package to compute the gradient.

```
<< Calculus`VectorAnalysis`
```

```
SetCoordinates[Cartesian[x, y, z]]
```

```
Cartesian[x, y, z]
```

```
field[x_, y_, z_, q_] = -Grad[phi[x, y, z, q]]//
  Simplify
```

$$\left\{ q\,x \left(-\frac{1}{(x^2 + y^2 + (-1 + z)^2)^{3/2}} + \frac{1}{(x^2 + y^2 + (1 + z)^2)^{3/2}} \right), \right.$$

$$q\,y \left(-\frac{1}{(x^2 + y^2 + (-1 + z)^2)^{3/2}} + \frac{1}{(x^2 + y^2 + (1 + z)^2)^{3/2}} \right),$$

$$\left. -\frac{q\,(-1 + z)}{(x^2 + y^2 + (-1 + z)^2)^{3/2}} + \frac{q\,(1 + z)}{(x^2 + y^2 + (1 + z)^2)^{3/2}} \right\}$$

```
conplot = ContourPlot[phi[x, 0, z,1] , {x, -1, 1},
    {z, 0, 2}, Contours -> 40,
    ColorFunction -> Hue, PlotPoints -> 50];
```

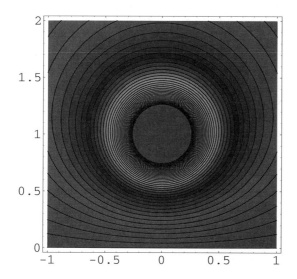

Our electric field is a three-dimensional vector field. To use **PlotVectorField**, we need a two-dimensional vector field. We can use the command **Drop** to achieve that. That is, we need to "drop" the *y*-component of the field since we are restricting attention to the *x-z* plane.

? Drop

```
Drop[list, n] gives list with its first n elements dropped. Drop[
    list, -n] gives list with its last n elements dropped.
    Drop[list, {n}] gives list with its nth element dropped.
    Drop[list, {m, n}] gives list with elements m through n dropped.
```

<< Graphics`PlotField`

field2D[x_, z_] = Drop[field[x, 0, z, 1],{2}]

$$\left\{ x\left(-\frac{1}{\left(x^2 + (-1 + z)^2 \right)^{3/2}} + \frac{1}{\left(x^2 + (1 + z)^2 \right)^{3/2}} \right), \right.$$
$$\left. -\frac{-1 + z}{\left(x^2 + (-1 + z)^2 \right)^{3/2}} + \frac{1 + z}{\left(x^2 + (1 + z)^2 \right)^{3/2}} \right\}$$

Since the field blows up at (0, 1), we'll set it equal to (0, 0) just at that point to get rid of the singularity.

```
goodfield2D[x_,  z_] = If[x^2 + (z-1)^2 > 0,
 field2D[x,  z], {0, 0}]
```

If $[x^2 + (-1 + z)^2 > 0$, field2D$[x, z]$, $\{0, 0\}]$

```
fieldplot2D =
 PlotVectorField[Evaluate[goodfield2D[x,  z]],
 {x, -1, 1},  {z, 0, 2}];
```

Finally, we show the potential and the electric field together.

`Show[conplot, fieldplot2D];`

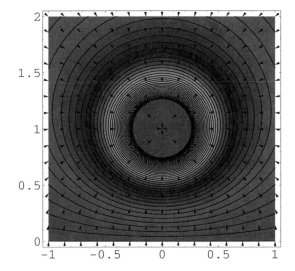

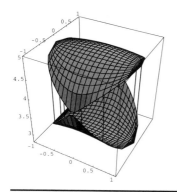

Index

The index uses the same conventions for fonts that are used throughout the book. *Mathematica* commands, such as **DSolve**, are printed in typewriter boldface. Menu options, such as `Cell`, are printed in a monospaced typewriter font. Everything else is printed in a standard font.

Multivariable Calculus and *Mathematica*®

With Applications to Geometry and Physics

Since this field is fast-moving, we expect updates and changes to occur that might necessitate sending you the most current pertinent information by paper, electronic media, or both, regarding *Multivariable Calculus and Mathematica*®. Therefore, in order to not miss out on receiving your important update information, please fill out this card and return it to us promptly. Thank you.

Name: —————————————————————————

Title: —————————————————————————

Company: ————————————————————————

Address: ————————————————————————

City: ——————————————— State: —— Zip: ———

E-mail: ——————————————————————————

Areas of Interest/Technical Expertise: ———————————

Comments on this Publication: ——————————————

—————————————————————————————

—————————————————————————————

❑ Please check this box to indicate that we may use your comments in our promotion and advertising for this publication.

Purchased from: ——————————————————

Date of Purchase: ——————————————————

❑ Please add me to your mailing list to receive updated information on *Multivariable Calculus and Mathematica*® and other TELOS publications.

I have a(n)   ❑ IBM compatible   ❑ Macintosh   ❑ Unix   ❑ Other

Designate specific model: ———————————————

TELOS® THE ELECTRONIC LIBRARY OF SCIENCE

BUSINESS REPLY MAIL
FIRST-CLASS MAIL   PERMIT NO. 5863   NEW YORK, NY

POSTAGE WILL BE PAID BY ADDRESSEE

THE
ELECTRONIC
LIBRARY
OF
SCIENCE

TELOS PROMOTION
SPRINGER-VERLAG NEW YORK, INC.
ATTN: J. Roth
175 FIFTH AVENUE
NEW YORK NY 10160-0266